Springer Tracts in Modern Physics
Volume 169

Now also Available Online

Starting with Volume 163, Springer Tracts in Modern Physics is part of the Springer LINK service. For all customers with standing orders for Springer Tracts in Modern Physics we offer the full text in electronic form via LINK free of charge. Please contact your librarian who can receive a password for free access to the full articles by registration at:

http://link.springer.de/series/stmp/reg_form.htm

If you do not have a standing order you can nevertheless browse through the table of contents of the volumes and the abstracts of each article at:

http://link.springer.de/series/stmp/

There you will also find more information about the series.

Springer
Berlin
Heidelberg
New York
Barcelona
Hong Kong
London
Milan
Paris
Singapore
Tokyo

Physics and Astronomy   ONLINE LIBRARY

http://www.springer.de/phys/

Springer Tracts in Modern Physics

Springer Tracts in Modern Physics provides comprehensive and critical reviews of topics of current interest in physics. The following fields are emphasized: elementary particle physics, solid-state physics, complex systems, and fundamental astrophysics.

Suitable reviews of other fields can also be accepted. The editors encourage prospective authors to correspond with them in advance of submitting an article. For reviews of topics belonging to the above mentioned fields, they should address the responsible editor, otherwise the managing editor. See also http://www.springer.de/phys/books/stmp.html

Managing Editor

Gerhard Höhler

Institut für Theoretische Teilchenphysik
Universität Karlsruhe
Postfach 69 80
76128 Karlsruhe, Germany
Phone: +49 (7 21) 6 08 33 75
Fax: +49 (7 21) 37 07 26
Email: gerhard.hoehler@physik.uni-karlsruhe.de
http://www-ttp.physik.uni-karlsruhe.de/

Elementary Particle Physics, Editors

Johann H. Kühn

Institut für Theoretische Teilchenphysik
Universität Karlsruhe
Postfach 69 80
76128 Karlsruhe, Germany
Phone: +49 (7 21) 6 08 33 72
Fax: +49 (7 21) 37 07 26
Email: johann.kuehn@physik.uni-karlsruhe.de
http://www-ttp.physik.uni-karlsruhe.de/~jk

Thomas Müller

Institut für Experimentelle Kernphysik
Fakultät für Physik
Universität Karlsruhe
Postfach 69 80
76128 Karlsruhe, Germany
Phone: +49 (7 21) 6 08 35 24
Fax: +49 (7 21) 6 07 26 21
Email: thomas.muller@physik.uni-karlsruhe.de
http://www-ekp.physik.uni-karlsruhe.de

Fundamental Astrophysics, Editor

Joachim Trümper

Max-Planck-Institut für Extraterrestrische Physik
Postfach 16 03
85740 Garching, Germany
Phone: +49 (89) 32 99 35 59
Fax: +49 (89) 32 99 35 69
Email: jtrumper@mpe-garching.mpg.de
http://www.mpe-garching.mpg.de/index.html

Solid-State Physics, Editors

Andrei Ruckenstein
Editor for The Americas

Department of Physics and Astronomy
Rutgers, The State University of New Jersey
136 Frelinghuysen Road
Piscataway, NJ 08854-8019, USA
Phone: +1 (732) 445 43 29
Fax: +1 (732) 445-43 43
Email: andreir@physics.rutgers.edu
http://www.physics.rutgers.edu/people/pips/Ruckenstein.html

Peter Wölfle

Institut für Theorie der Kondensierten Materie
Universität Karlsruhe
Postfach 69 80
76128 Karlsruhe, Germany
Phone: +49 (7 21) 6 08 35 90
Fax: +49 (7 21) 69 81 50
Email: woelfle@tkm.physik.uni-karlsruhe.de
http://www-tkm.physik.uni-karlsruhe.de

Complex Systems, Editor

Frank Steiner

Abteilung Theoretische Physik
Universität Ulm
Albert-Einstein-Allee 11
89069 Ulm, Germany
Phone: +49 (7 31) 5 02 29 10
Fax: +49 (7 31) 5 02 29 24
Email: steiner@physik.uni-ulm.de
http://www.physik.uni-ulm.de/theo/theophys.html

Y. Yamamoto   T. Tassone   H. Cao

Semiconductor Cavity Quantum Electrodynamics

With 67 Figures

 Springer

Professor Yoshihisa Yamamoto

Stanford University
Edward L. Ginzton Laboratory
Stanford, CA 94305-4085,USA
E-mail: yamamoto@loki.stanford.edu

Professor Francesco Tassone

SISSA
v. Beirut 4
34014 Trieste, Italy
E-mail: tassone@sissa.it

Professor Hui Cao

Northwestern University
Dept. of Physics and Astronomy
2145 Sheridan Road
IL 60208 Evanston, USA
E-mail: huicao@leland.stanford.edu

Library of Congress Cataloging-in-Publication Data.

Yamamoto, Yoshihisa. Semiconductor cavity quantum electrodynamics / Y. Yamamoto, F. Tassone, H. Cao. p.cm.
– (Springer tracts in modern Physics, ISSN 0081-3869; v. 169) Includes bibliographical references and index.
ISBN 3-540-67520-5 (alk. paper). 1. Quantum electrodynamics. 2. Quantum electronics. 3. Semiconductors–Optical
properties. 4. Solid state physics. I. Tassone, F. (Francesco), 1967– II. Cao, H. (Hui), 1968– III. Title. IV. Springer
tracts in modern physics; 169.   QC680.Y36 2000   530.14'33–dc21   00-059483

Physics and Astronomy Classification Scheme (PACS):
42.50.-P, 42.55.Sa, 78.20.-e, 78.45.+h, 78.47.+p

ISSN print edition: 0081-3869
ISSN electronic edition: 1615-0430
ISBN 3-540-67520-5 Springer-Verlag Berlin Heidelberg New York

Springer-Verlag Berlin Heidelberg New York
a member of BertelsmannSpringer Science+Business Media GmbH

© Springer-Verlag Berlin Heidelberg 2000
Printed in Germany

Typesetting: Data conversion by Steingraeber Satztechnik GmbH, Heidelberg using a Springer LaTeX macro package
Cover design: *design & production* GmbH, Heidelberg

Printed on acid-free paper   SPIN: 10676950   56/3141/tr   5 4 3 2 1 0

Preface

This book reviews the recent experimental and theoretical progress in semiconductor cavity quantum electrodynamics (QED) in the strong-coupling regime. The contents of the book are based on experimental and theoretical results obtained by the authors' research group at Stanford University, California, but we also try to refer to the relevant work of other groups around the world as much as possible. As is well known today, spontaneous emission is not an immutable property of an atom but is a consequence of atom–vacuum field coupling. The study of atom–vacuum field interactions in cavities and of modified spontaneous emission constitutes the growing field of atomic cavity QED. The basic concept can be traced back to a pioneering paper by E. M. Purcell in 1946. The research on semiconductor (or excitonic) cavity QED started in the late 1980s. The early experiments belong to the weak-coupling regime, for which the spontaneous emission is still an irreversible process. However, a modified spontaneous-emission decay rate and altered radiation pattern have been clearly demonstrated. It is expected that the quantum efficiency, response time, and intensity noise of a semiconductor light emitter can be improved by cavity QED effects. Our own work on such semiconductor cavity QED in the weak-coupling regime is summarized in review articles published ten years ago [49, 171]. The research on semiconductor cavity QED entered into the strong-coupling regime in the early 1990s with the pioneering experimental work of C. Weisbuch and coworkers [50]. The quantum well (QW) exciton and the cavity exchange energy coherently, and thus spontaneous emission becomes a partially reversible process. This rapidly growing field has produced numerous interesting results in fundamental science in the past decade. However, the practical applications of semiconductor cavity QED in the strong-coupling regime have yet to be identified. We have tried to address the future prospects and practical applications of semiconductor cavity QED in a high-Q regime in this book.

Chapter 1 introduces the basic concepts of atomic and semiconductor cavity QED, such as modified spontaneous emission in a cavity, Rabi oscillations, the one-atom maser, excitons, the excitonic oscillator strength, bulk exciton polaritons, QW excitons, semiconductor microcavities, and exciton–photon coupling. Chapter 2 gives an overview of the theory and experiments on microcavity exciton polaritons. The transition from the strong-coupling to the weak-coupling regime is closely examined with both time-domain and frequency-domain measurements. Chapter 3 describes another type of nonlinearity, the biexcitonic effect, of microcavity exciton polaritons. The concept of a dressed exciton is introduced to describe weakly interacting exciton polaritons. Chapter 4 discusses the direct creation of excitons and microcavity polaritons by resonant tunneling of electrons into the QW exciton and polariton states. This tunneling process is fundamentally different from the conventional resonant tunneling process, because the final state of the tunneling is a bosonic exciton or exciton–polariton state instead of a fermionic electron state. If there is a finite population in the final state, the resonant-tunneling rate is enhanced by the bosonic final-state stimulation, while in the standard tunneling process with final-state occupancy, the resonant-tunneling rate is suppressed by the Pauli exclusion principle. Chapter 5 introduces the concept of the exciton laser or boser, that is, stimulated emission of massive bosonic particles of excitons or exciton polaritons in a semiconductor microcavity. The exciton laser is different from the Bose–Einstein condensation of excitons. The former is realized with a nonequilibrium (inverted) reservoir of excitons, but the latter is the property of an equilibrium, closed system. We address the technological difficulties of achieving an exciton laser and the inherent competition between exciton lasing and photon lasing in the same system. Chapter 6 describes one promising way to realize an exciton laser, which is based on the formation of an inverted exciton reservoir by the so-called bottleneck effect, with exciton–exciton collisions, as a gain mechanism. A rate equation analysis and preliminary experimental results are presented. Chapter 7, finally, analyzes the quantum statistical properties of an exciton laser.

We would like to thank our colleagues and friends who have contributed so much to the material of the book: in particular, Gunner Björk, Joseph Jacobson, Stanley Pau, Gleb Klimovitch, Susumu Machida, Shudong Jiang, Yoshihiro Takiguchi, Eiichi Hanamura, Atac Imamoglu, Hilin Wang, Ross Stanley, and Paola Pellandini. We also would like to thank Mayumi Hakkaku for her patient and careful word-processing of the manuscript.

Stanford, April, 2000   *Yoshihisa Yamamoto*
                                                        Francesco Tassone
                                                        Hui Cao

Contents

1. Introduction

1.1 Atomic Cavity QED

For a long time, spontaneous emission was regarded as an inherent property of matter. However, this view overlooks the fact that spontaneous emission is not an immutable property of an atom, but a consequence of atom–vacuum field coupling. The most distinctive feature of spontaneous emission, irreversibility, comes about because an infinity of vacuum states is available to the radiated photon. If these states are modified – for instance, by placing the excited atom between mirrors or in a cavity – spontaneous emission can be greatly inhibited or enhanced.

The extensive studies of atom–vacuum interactions in cavities constitute the rapidly growing field of "cavity quantum electrodynamics" (cavity QED) [1]. In this chapter, a brief review of atomic cavity QED will be presented.

1.1.1 Emission in Free Space

We start with spontaneous emission in free space. Consider a one-electron atom with two electronic levels e and f separated by an energy interval $E_e - E_f = \hbar\omega$. Spontaneous emission appears as a jump of the electron from level e to f accompanied by the emission of a photon. This process can be understood as resulting from the coupling of the atomic electron to the electromagnetic field in its vacuum state.

A radiation field in space is usually described in terms of an infinite set of harmonic oscillators, one for each mode of the radiation. The levels of each oscillator correspond to states with $0, 1, 2, ..., n$ photons of energy $\hbar\omega$. In its ground state, each oscillator has a zero-point energy $\hbar\omega/2$ associated with its quantum fluctuations.

The rms vacuum electric-field amplitude E_{vac} in a mode of frequency ω is

$$E_{\text{vac}} = \left(\frac{\hbar\omega}{2\epsilon_0 V} \right)^{1/2} , \tag{1.1}$$

where ϵ_0 is the permittivity of free space, V is the size of an arbitrary quantization volume, and the units are SI. The coupling of an atom to a field mode is described by the frequency

$$\Omega_{ef} = \frac{D_{ef} E_{\mathrm{vac}}}{\hbar} \;, \tag{1.2}$$

where D_{ef} is the matrix element of the electric dipole of the atom between the two levels. Ω_{ef}, which is often referred to as the Rabi frequency of the vacuum, is the frequency at which the atom and the field would exchange energy if there were only a single mode of the field. An essential feature of spontaneous emission in free space is that an atom can radiate into any mode that satisfies the conservation of energy and momentum. The time of emission and the particular mode in which the photon is observed are random variables.

The probability Γ_0 of photon emission per unit time, more familiarly called the Einstein A coefficient, is proportional the square of the frequency Ω_{ef} and to the mode density $\rho_0(\omega)$ (the number of modes available per unit frequency interval). The mode density is given by the expression $\rho_0(\omega) = \omega^2 V/\pi^2 c^3$, where it is assumed that the quantization volume V is large compared with λ^3, or $(2\pi c/\omega)^3$. The probability Γ_0 is given by the Fermi's golden rule:

$$\Gamma_0 = 2\pi \Omega_{ef}^2 \frac{\rho_0(\omega)}{3} = \frac{\omega^3}{3\pi\hbar c^3} \frac{|D_{ef}|^2}{\epsilon_0} \;. \tag{1.3}$$

The probability $P_e(t)$ of finding an atom still excited at time t after its preparation in state e at $t = 0$ is

$$P_e(t) = \mathrm{e}^{-\Gamma_0 t} \;. \tag{1.4}$$

Such an exponential decay law describes an irreversible process that leads the atom irrevocably to its ground state. The source of irreversibility is the continuum of field modes resonantly coupled to the atom. The vacuum field acts as a gigantic reservoir in which the atomic excitation decays away.

1.1.2 Spontaneous Emission in a Cavity

The mode structure of the vacuum field is dramatically altered in a cavity whose size is comparable to the wavelength. Consider an atom confined between two plane parallel mirrors with separation d. For an electric field polarized parallel to the mirrors, no mode exists unless $\lambda < 2d$. As λ is increased, the lowest-order mode cuts off abruptly when λ exceeds $2d$. As a result, an excited atom whose radiation arises from an electric dipole moment oscillating parallel to the mirrors becomes infinitely long-lived when $\lambda > 2d$.

Just as a cavity below cutoff suppresses vacuum fluctuations, a resonant cavity enhances them. How an atom behaves in a resonant cavity depends on the ratio of the vacuum Rabi frequency Ω_{ef} to the cavity bandwidth $\Delta\omega_{\mathrm{c}}$. The cavity bandwidth is most conveniently described by the quality factor Q, which is given by $\omega/\Delta\omega_{\mathrm{c}}$. The reciprocal of $\Delta\omega_{\mathrm{c}}$ is the density of modes "seen" by the atom in the cavity; alternatively, it is the lifetime of a photon in the cavity.

In a low-Q cavity, the emitted photon is damped rapidly and an atom undergoes radiative decay much as it does in free space, though at an enhanced rate. The radiation rate in a cavity of volume V is

$$\Gamma_{\text{cav}} \cong \Gamma_0 \frac{Q\lambda^3}{V} \,. \tag{1.5}$$

Compared with the rate in free space Γ_0, the emission rate Γ_{cav} is increased by the ratio $Q\lambda^3/V$, which can be large.

The modification of a spontaneous-emission decay rate by a cavity wall was first predicted by Purcell in 1946 [2], and experimentally demonstrated by Drexhage in the late 1960s [3]. In the past ten years, both enhanced and inhibited spontaneous emission from atoms placed inside cavities has been observed by several groups, in microwave and optical regimes [4–8].

In addition to changing the radiation rate of atoms, atom–cavity coupling also induces energy shifts [9, 10]. This phenomenon can be understood as a frequency-pulling effect that occurs when an atomic oscillator interacts with a reactive cavity. Quantum mechanically, it can be analyzed in terms of the effect of the cavity walls on the virtual-photon exchange processes. The boundaries alter the modes of the vacuum field around the atom and affect not only real-photon emission (the spontaneous-emission rate), but also the virtual process responsible for radiative energy shifts.

1.1.3 Rabi Oscillations

The regime of very high Q, where $\Omega_{ef}/\Delta\omega_c > 1$, manifests totally new behavior. The radiation remains in the cavity so long that there is a high probability it will be reabsorbed by the atom before it dissipates. Spontaneous emission becomes reversible, as the atom and the field exchange excitation at the rate Ω_{ef}. Such behavior is a well-known feature of the interaction of an atom with a classical monochromatic field. These so called "Rabi oscillations" are familiar in nuclear magnetic resonance and optical-transient experiments. In cavity QED, however, the atom couples to its own one-photon field without an external applied field. Thus this effect is called "vacuum Rabi oscillation".

When there are N atoms in the cavity, the vacuum Rabi oscillation frequency is increased, i.e.

$$\Omega_{\text{R}}(N) = \Omega_{ef}\sqrt{N} \,. \tag{1.6}$$

Therefore, the first observation of vacuum Rabi oscillation was made with a large number of atoms in a microwave cavity [11]. Subsequent improvement of the cavity Q value has led to the observation of vacuum Rabi oscillation for a few atoms [12] and, finally, for a single atom [13].

On the other hand, when there are n photons in the cavity, the vacuum Rabi oscillation frequency becomes

$$\Omega_{\text{R}}(n) = \Omega_{ef}\sqrt{n+1} \,. \tag{1.7}$$

Rempe et al. first observed the Rabi oscillations induced by a small thermal field in a cavity [14]. The probability $P_e(t)$ of finding the atom in the state e at time t is

$$P_e(t) = \sum_n p(n) \cos^2 \left(\frac{1}{2} \Omega_{ef} \sqrt{n+1}\, t \right) , \tag{1.8}$$

where $p(n)$ is the probability of n photons in the cavity. After a few oscillation periods, the various terms corresponding to different values of n interfere destructively, leading to a collapse of the temporal oscillation of $P_e(t)$. Some time later, the terms interfere constructively and the beating revives. This phenomenon is called "the quantum collapse and revival" of the Rabi oscillations.

1.1.4 One-Atom Masers/Lasers

In the nonlinear regime, lasing oscillation has been achieved with exceedingly small numbers of atoms and photons in very-high-Q cavities [15–19]. This is a new kind of maser/laser: there is never more than a single atom in the resonator – in fact, most of the time the cavity is empty.

Consider a stream of atoms prepared in state e passing through a cavity. If the rate of atoms crossing the cavity exceeds the cavity damping rate ω/Q, a photon released by an atom is stored long enough to interact with the next atom. The atom–field coupling becomes stronger and stronger as the field builds up, eventually evolving into a steady state. After the first atom has crossed the resonator, the state of the atom–field system is a linear superposition of states $|e, 0\rangle$ (atom in state e with 0 photons in the cavity) and $|f, 1\rangle$ (atom in state f with 1 photon in the cavity), i.e.,

$$|\Psi_1\rangle = \cos \left(\frac{\Omega_{ef}\, t_{\text{int}}}{2} \right) |e, 0\rangle + \text{sim} \left(\frac{\Omega_{ef}\, t_{\text{int}}}{2} \right) |f, 1\rangle . \tag{1.9}$$

If the first atom is found in state $|e\rangle$ after passing through the resonator, the cavity field contains no photons. The next atom will then interact with an empty cavity, just like the first atom. However, if the first atom is found in state $|f\rangle$ after passing through the resonator, one photon appears in the cavity. The next atom will then undergo a different evolution, i.e. the state of the system after the second atom has left the cavity will be

$$|\Psi_1\rangle = \cos(\Omega_{ef} \sqrt{2}\, t_{\text{int}}/2)|e, 0\rangle + \sin(\Omega_{ef} \sqrt{2}\, t_{\text{int}}/2)|f, 1\rangle . \tag{1.10}$$

Continuous monitoring of the system would reveal a kind of random walk of the photon number: each time an atom is detected in state $|f\rangle$, the cavity field has gained one photon. The process is essentially random because the probability for an atom to flip from state e to f is governed by quantum mechanical chance. It is a random process with memory, however, since the probability law for each step depends on the outcome of the previous steps. A steady state is reached if there exists a photon number n_0 such that the quantity $(\Omega_{ef}\sqrt{n_0+1}\,t_{\rm int})/2$ is an integer multiple of π. Then the photon number remains trapped at the value n_0 because each subsequent atom will leave the cavity in state e. Such a radiation state is highly nonclassical, for the field has a precisely defined energy but a completely random phase. It has no amplitude fluctuations, whereas an ordinary electromagnetic field has quantum-limited intensity fluctuations proportional to $\sqrt{\bar{n}}$. Such nonclassical states, i.e. amplitude-squeezed states, have been observed in a high-Q microcavity [20].

1.2 Excitons

1.2.1 Excitons Versus Electron–Hole Pairs

The calculation of semiconductor band structures is based on a one-electron approximation. However, when one electron is excited from the filled valence band to the empty conduction band, a pair excitation is created which consists of a missing valence band electron (i.e. a hole) and a conduction band electron. A bound state of a conduction band electron and a valence band hole is called an exciton, while an unbound state of an electron and a hole is called a free electron–hole pair. Excitons in semiconductors are usually shallow, i.e. their radius is much larger than the interatomic spacing: as such they can be described by a two-particle effective-mass equation. The opposite extreme, i.e. that of tightly bound (or Frenkel) excitons, is appropriate to molecular crystals. In this chapter a brief introduction to shallow (or Wannier–Mott) excitons in semiconductors will be presented.

In a direct two-band semiconductor, the Hamiltonian of the system can be written as [21]

$$
\begin{aligned}
H = \sum E_{\rm e}(k)\,a_{\boldsymbol{k}}^{\dagger}a_{\boldsymbol{k}} + \sum E_{\rm h}(k)\,b_{\boldsymbol{k}}^{\dagger}b_{\boldsymbol{k}} \\
+\frac{1}{2}\sum V_{\boldsymbol{k}_1\boldsymbol{k}_2\boldsymbol{k}_3\boldsymbol{k}_4}^{\rm cccc}\,a_{\boldsymbol{k}_1}^{\dagger}a_{\boldsymbol{k}_2}^{\dagger}a_{\boldsymbol{k}_3}a_{\boldsymbol{k}_4} \\
+\frac{1}{2}\sum V_{-\boldsymbol{k}_1-\boldsymbol{k}_2-\boldsymbol{k}_3-\boldsymbol{k}_4}^{\rm vvvv}\,b_{\boldsymbol{k}_1}^{\dagger}b_{\boldsymbol{k}_2}^{\dagger}b_{\boldsymbol{k}_3}b_{\boldsymbol{k}_4} \\
-\sum (V_{\boldsymbol{k}_1\boldsymbol{k}_3\boldsymbol{k}_2\boldsymbol{k}_4}^{\rm cvvc}-V_{\boldsymbol{k}_1\boldsymbol{k}_3\boldsymbol{k}_4\boldsymbol{k}_2}^{\rm cvcv})\,a_{\boldsymbol{k}_1}^{\dagger}b_{\boldsymbol{k}_2}^{\dagger}b_{\boldsymbol{k}_3}a_{\boldsymbol{k}_4}\,,
\end{aligned}
\tag{1.11}
$$

where $a_{\boldsymbol{k}}$ and $b_{\boldsymbol{k}}$ are the Fermi annihilation operators for electrons and holes in the conduction and valence band, respectively, $E_{\rm e}(k)=E_{\rm g}+\hbar^2k^2/2m_{\rm e}$, $E_{\rm h}(k)=\hbar^2k^2/2m_{\rm h}$, $E_{\rm g}$ is the bandgap, $m_{\rm e}$ and $m_{\rm h}$ are the effective masses of the conduction band electron and the valence band hole, respectively, $V_{\boldsymbol{k}_1\boldsymbol{k}_2\boldsymbol{k}_3\boldsymbol{k}_4}^{ijlm}=\langle\boldsymbol{k}_1 i,\boldsymbol{k}_2 j|V|\boldsymbol{k}_3 l,\boldsymbol{k}_2 m\rangle$, and V includes the Coulomb and exchange interaction between electrons and holes.

The next step is to solve for the eigenfunctions of the general electron–hole pair state $|p\rangle=\sum_{k,k'}C_{\boldsymbol{k}\boldsymbol{k}'}\,a_{\boldsymbol{k}}^{\dagger}b_{\boldsymbol{k}'}^{\dagger}|0\rangle$. $|0\rangle$ is the vacuum state for pair generation, i.e. an empty conduction band and a full valence band. From $H|0\rangle = E|0\rangle$, we get an equation for the amplitude $C_{k\,k'}$

$$[E_e(k) + E_h(k') - E]\,C_{k\,k'} - \sum_{l,\,l'}(V^{cvvc}_{k-l'-k'l} - V^{cvcv}_{k-l'l-k'})\,C_{l\,l'} = 0 \,. \quad (1.12)$$

For low-lying electronic excitations, the plane wave factors and the Coulomb potential are slowly varying functions, which change very little in one unit cell. In the effective-mass approximation, the Coulomb and exchange potentials in real space can be significantly simplified, leading to the Wannier equation for an exciton [22]:

$$H_{ex}\Phi(x_e,\,x_h) = E\Phi(x_e,\,x_h) \,, \quad (1.13)$$

with

$$H_{ex} = -\frac{\hbar^2}{2m_e}\Delta_e - \frac{\hbar^2}{2m_e}\Delta_h + E_g - \frac{e^2}{\epsilon|x_e - x_h|} \,, \quad (1.14)$$

where the Schrödinger two-particle wavefunction is related to the amplitudes $C_{k\,k'}$ by

$$\Phi(x_e,\,x_h) = \sum C_{k\,k'}e^{ik\cdot x_e + ik'\cdot x_h} \,. \quad (1.15)$$

By introducing the relative coordinate $r = x_e - x_h$ and the center-of-mass coordinate $R = (m_e\,x_e + m_h\,x_h)/M$, where $M = m_e + m_h$, the center-of-mass motion can be separated by means of

$$\Phi(x_e,\,x_h) = \varphi_n(r)\frac{1}{\sqrt{V}}e^{iK\cdot R} \,, \quad (1.16)$$

where K corresponds to the center-of-mass momentum. The resulting equation of relative motion is

$$\left(-\frac{\hbar^2}{2\mu}\Delta_r - \frac{e^2}{\epsilon\,r} + E_{b,n}\right)\varphi_n(r) = 0 \,, \quad (1.17)$$

where the reduced mass $\mu = m_e\,m_h/(m_e + m_h)$.

The total energy of the pair is given by

$$E(K,n) = E_g - E_{b,n} + \frac{\hbar^2 K^2}{2M} \,, \quad (1.18)$$

with the binding energy of the pair is

$$E_{b,n} = \frac{R^*}{n^2} = \frac{\hbar^2}{2\mu(a_B^*)^2\,n^2} \,, \quad (1.19)$$

where a_B is the exciton Bohr radius

$$a_B = \frac{\hbar^2\epsilon_0}{e^2\mu} \,, \quad (1.20)$$

and R^* is the Rydberg of the exciton

$$R^* = \frac{\mu \, e^4}{2\epsilon_0^2 \, \hbar^2} = \frac{\hbar^2}{2\mu(a_B^*)^2} \, . \tag{1.21}$$

We can now introduce an exciton operator $C_{\boldsymbol{k},n}$:

$$C_{\boldsymbol{k},n}^\dagger = \sum_{\boldsymbol{k},\boldsymbol{k}'} \delta_{\boldsymbol{K},\boldsymbol{k}+\boldsymbol{k}'} \, \varphi_n(\boldsymbol{l}) \, a_{\boldsymbol{k}}^\dagger b_{\boldsymbol{k}'}^\dagger \, , \tag{1.22}$$

with $\boldsymbol{l} = (m_{\mathrm{h}}\boldsymbol{k} - m_{\mathrm{e}}\boldsymbol{k}')/M$, and $\varphi_n(\boldsymbol{l}) = \int \mathrm{d}^3\boldsymbol{x} \, \varphi_n(\boldsymbol{x}) \exp(-\mathrm{i}\boldsymbol{l} \cdot \boldsymbol{x})$.

The commutation relation of the exciton operators can be derived:

$$[C_{\boldsymbol{k}'n'}, C_{\boldsymbol{k}n}] = \delta_{\boldsymbol{K},\boldsymbol{K}'} \, \delta_{n,n'} - \sum \varphi_{n'}^*(\alpha_1\boldsymbol{K}' - \boldsymbol{l}) \, \varphi_n(\alpha_1\boldsymbol{K} - \boldsymbol{l}) \, a_{\boldsymbol{K}-\boldsymbol{l}}^\dagger \, a_{\boldsymbol{K}'-\boldsymbol{l}}$$

$$- \sum \varphi_{n'}^*(\boldsymbol{l} - \alpha_2\boldsymbol{K}') \, \varphi_n(\boldsymbol{l} - \alpha_2\boldsymbol{K}) \, b_{\boldsymbol{K}-\boldsymbol{l}}^\dagger \, b_{\boldsymbol{K}'-\boldsymbol{l}}$$

$$= \delta_{\boldsymbol{K},\boldsymbol{K}'} \, \delta_{n,n'} + O(na_B^3) \, , \tag{1.23}$$

where $\alpha_1 = m_{\mathrm{h}}/M$ and $\alpha_2 = m_{\mathrm{e}}/M$.

Hence, excitons can be considered to be approximately bosons as long as the density of electronic excitations is small compared with a_B^{-3}, or in other words as long as the mean distance between two excitons is much larger than the extension of an exciton.

1.2.2 Excitonic Oscillator Strength

Although an exciton has a similar internal structure to a hydrogen atom, a Wannier exciton has much larger Bohr radius and much smaller binding energy than a hydrogen atom, owing to the small effective mass of electrons and the large dielectric constant in semiconductors. For instance, in the ground (1s) state, the electron and hole in a Wannier exciton are much more loosely bound than the electron and proton in a hydrogen atom. On the other hand, since the exciton's wavefunction is extended over many lattice sites of the crystal, the exciton collects the dipole oscillator strength of many atoms. More rigorously, the excitonic oscillator strength can be derived as follows [23]:

$$f_{\mathrm{exciton}} = \frac{2}{m_0\hbar\omega} \, |\langle f|\hat{\boldsymbol{e}} \cdot \sum \boldsymbol{p}_i|i\rangle|^2$$

$$= \frac{2}{m_0\hbar\omega} \, |\langle u_{\mathrm{v}}|\hat{\boldsymbol{e}} \cdot \boldsymbol{p}|u_{\mathrm{c}}\rangle|^2 \, V \, 2 \, |\psi_n(r = 0)|^2 \, \delta_{\boldsymbol{k}_{\mathrm{ex}},0} \, , \tag{1.24}$$

where the initial ($|i\rangle$) and final ($|f\rangle$) states refer to the crystal ground state and the exciton state, u_{v} and u_{c} represent the Bloch wavefunctions of the valence band and conduction band, V is the crystal volume, $\psi_n(r)$ is the exciton envelope function, and m_0 is free-electron mass.

For the 1s state,

$$|\psi_{1s}(r = 0)|^2 = \frac{1}{\pi(a_B^*)^3} .$$ (1.25)

Therefore,

$$f_{\text{exciton}} = f_{\text{atom}} \times \frac{V}{\pi(a_B^*)^3} ,$$ (1.26)

where f_{atom} is the oscillator strength for a delocalized electron–hole pair. This indicates that the strong electron–hole correlation in an exciton enhances its oscillator strength by a factor of $V/\pi(a_B^*)^3$ as compared with the uncorrelated electron–hole pair.

Equation (1.26) shows that the dimensionless oscillator strength is proportional to the crystal volume. This is due to the fact that the exciton center-of-mass wavefunction extends over the whole crystal. The meaningful quantity is the oscillator strength per unit volume, which is related to the absorption coefficient $\alpha(\omega)$ by

$$\int \alpha(\omega)\,d\omega = \frac{2\pi^2 e^2}{n\,m_0\,c}\frac{f}{V} .$$ (1.27)

The excitonic effect introduces discrete absorption lines below the band gap. Above the band gap, excitonic effects give a correlation in the electron and hole of the unbound electron–hole pair, leading to an enhancement factor (often called the Sommerfeld factor) of the absorption coefficient:

$$S(\omega) = \frac{2\pi x}{1 - \exp(-2\pi x)} , \quad x = \left(\frac{R^*}{\hbar\omega - E_g}\right)^{1/2} .$$ (1.28)

Figure 1.1 shows a schematic plot of the absorption coefficient with and without excitonic effects.

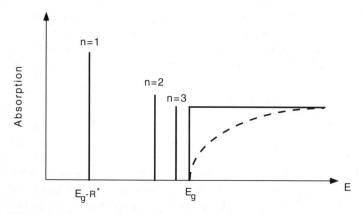

Fig. 1.1. Schematic picture of the absorption coefficient close to the interband absorption edge of 3D crystals. *Dashed line*: without excitonic effect. *Solid line*: including excitonic effects

If there is no spin–orbit interaction, exciton states can be classified by total spin, and they are either singlet or triplet states. The oscillator strength is nonzero only for singlet states, since the radiation–matter interaction does not involve spin. The exchange interaction induces an energy splitting between singlet and triplet states. Furthermore, the dipole–dipole interaction leads to an energy splitting between longitudinal and transverse excitons, called the "LT splitting" [24]:

$$\Delta E_{\mathrm{LT}} = \frac{2\pi}{\epsilon}\,\frac{\hbar e^2}{m_0 \omega}\,\frac{f}{V}\,.\tag{1.29}$$

1.2.3 Bulk Exciton Polaritons

Polaritons are the normal modes of optically active excitons and a radiation field in solids [25]. In other words, the polariton involves a spatially coherent coupling of the exciton with the optical field [26]. In an infinite crystal, only those excitons at the crossing of the noninteracting exciton and photon dispersions can decay, owing to to the requirement of energy–momentum conservation in the radiation process (see Fig. 1.2). However, the exciton–photon coupling leads to an anticrossing in the polariton dispersions. Thus exciton polaritons in the bulk are stationary. Their radiative decay must be associated with phonons, defects, impurities, or crystal interfaces.

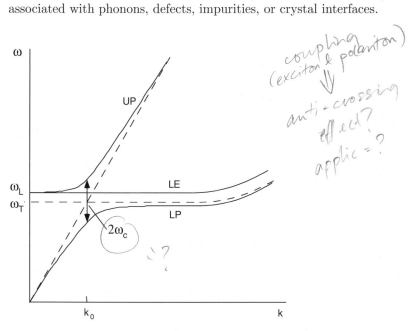

Fig. 1.2. Schematic representation of the dispersion of the upper and lower polaritons and of the longitudinal exciton (*solid lines*). The dispersion of the uncoupled photon and transverse exciton modes is also shown (*dashed lines*)

The semiclassical theory of exciton polaritons starts with the dielectric function:

$$\epsilon(\omega) = \epsilon_\infty + \frac{4\pi\beta\,\omega_0^2}{\omega_0^2 - \omega^2} \,, \tag{1.30}$$

where $\hbar\omega_0$ corresponds to the energy of the exciton, ϵ_∞ represents a frequency-independent contribution due to all other resonances in the crystal, and the polarizability β is related to the oscillator strength per unit volume by

$$\beta = \frac{e^2}{m_0\omega_0^2}\frac{f}{V} \,. \tag{1.31}$$

Maxwell's equations yield the following dispersion relations:

$$\frac{c^2 k^2}{\omega^2} = \epsilon(\omega) \tag{1.32}$$

for transverse modes, and

$$\epsilon(\omega) = 0 \tag{1.33}$$

for longitudinal modes. The dispersion relations are schematically illustrated in Fig. 1.2. The transverse modes show a lower and an upper branch, which anticross close to the wavevector of light. The longitudinal mode is a pure electrostatic solution, and its electrostatic shift is equal to the LT splitting:

$$\hbar\omega_{\mathrm{LT}} \simeq \frac{2\pi\beta}{\epsilon_\infty}\hbar\omega_0 \,. \tag{1.34}$$

A quantum theory of polaritons may be obtained by setting up a second-quantized Hamiltonian which describes excitons and photons, with their mutual interaction. The exciton–photon Hamiltonian can be derived either from a microscopic model for the exciton [25, 27] or from second quantization of Maxwell's equations plus the equation of motion for the excitonic polarization [23]:

$$H = \sum_k \left[\hbar v k \left(a_k^\dagger a_k + \frac{1}{2} \right) + \hbar\omega_k \left(a_k^\dagger a_k + \frac{1}{2} \right) \right.$$
$$\left. + iG_k(a_k^\dagger + a_{-k})(b_k - b_{-k}^\dagger) + D_k(a_k^\dagger + a_{-k})(a_k + a_{-k}^\dagger) \right] \,, \tag{1.35}$$

where $k = (\boldsymbol{k}, \sigma)$ is a combined index including the wavevector and polarization vector, a_k and b_k are Bose operators for the photon and exciton, respectively, $v = c/\sqrt{\epsilon}$ is the speed of light in the crystal, $\hbar\omega_k = \hbar\omega_0 + \hbar^2 k^2/2M$ is the exciton energy including spatial dispersion, and

$$G_k = \hbar\omega_0 \left(\frac{\pi\beta\,\omega_k}{v\,k\,\epsilon} \right)^{1/2} \,, \qquad D_k = \hbar\omega_0 \frac{\pi\beta\,\omega_k}{v\,k\,\epsilon} \,. \tag{1.36}$$

The normal modes of the above Hamiltonian can be found by a linear operator transformation [25]:

$$\begin{bmatrix} p_{k1} \\ p_{k2} \\ p^\dagger_{-k1} \\ p^\dagger_{-k2} \end{bmatrix} = \begin{bmatrix} W_1 & X_1 & Y_1 & Z_1 \\ W_2 & X_2 & Y_2 & Z_2 \\ Y^*_1 & Z^*_1 & W^*_1 & X^*_1 \\ Y_2* & Z^*_2 & W^*_2 & X^*_2 \end{bmatrix} \begin{bmatrix} a_k \\ b_k \\ a^\dagger_{-k} \\ b^\dagger_{-k} \end{bmatrix} . \tag{1.37}$$

The new operators p_{kl} $(l = 1, 2)$ are determined by the conditions that they satisfy Bose commutation relations, and that the Hamiltonian becomes diagonal, i.e.

$$[p_{kl}, H] = \hbar \Omega_l(k) \, p_{kl} . \tag{1.38}$$

This leads to the following secular equation for the eigenfrequencies $\Omega_l(k)$:

$$\Omega_l(k)^4 - \left(v^2 k^2 + \omega^2_k + \frac{4\pi\beta}{\epsilon_\infty} \omega^2_0 \right) \Omega_l(k)^2 + v^2 k^2 \omega^2_k = 0 . \tag{1.39}$$

The two solutions of the above equation correspond to the energies of the two bulk exciton–polariton modes.

The underlying physics of the bulk polaritons is that of two coupled harmonic oscillators for each wavevector $\boldsymbol{k}$; one corresponds to an exciton with center-of-mass momentum $\hbar\boldsymbol{k}$ and the other to an optical-field mode of wavevector $\boldsymbol{k}$. If we set one oscillator in motion when the other is initially at rest, there will be a periodic transfer of the energy of oscillation back and forth between the two oscillators. This periodic energy exchange between exciton and photon is the solid-state analog of the vacuum Rabi oscillation. However, the exciton polariton oscillation frequency is on the order of THz, which is much larger than the atom's vacuum Rabi frequency.

1.2.4 Quantum Well Excitons

When excitons are confined in a thin layer, two regimes can be distinguished according to the relative values of the layer thickness L and the exciton radius a_B. In the limit $L \gg a_B$, called the thin-film regime, the excitonic Rydberg R^* is much larger than the quantization energy $\hbar^2/(m_e L)$, and thus the exciton is only weakly perturbed by the confinement. The internal electron–hole wavefunction is undistorted, but the exciton center-of-mass motion is quantized [28,29]. In the limit $L \sim a_B$, called the quantum well (QW) regime, the excitonic Rydberg is smaller than the quantization energy of the subbands, and thus separate quantization of the electron and hole subbands occurs. The distortion of the internal exciton wavefunction due to a decrease of the average electron–hole separation leads to an increase of the binding energy and of the excitonic oscillator strength per unit area as the well thickness is reduced [30,31].

In the quantum well regime, the exciton wavefunction is represented in a basis consisting of products of conduction subbands $c(z_e)$ and valence subbands $v(z_h)$,

$$F(\rho, z_{\mathrm{e}}, z_{\mathrm{h}}) = \sum_{ijl} A_{ijl} \mathrm{e}^{-\alpha_l \rho} c_i(z_{\mathrm{e}}) v_j(z_{\mathrm{h}}) , \qquad (1.40)$$

where ρ is the in-plane relative coordinate of the electron and hole. In the limit $L \to 0$, if the barriers are taken to be of infinite height, the exciton binding energy tends to the value appropriate to the two-dimensional Coulomb problem, $R^*_{\mathrm{2D}} = 4R^*_{\mathrm{3D}}$ [32]. With a finite barrier height, the exciton binding energy reaches a maximum and then approaches the value corresponding to the barrier material as the well thickness decreases toward zero [33]. A quantitative determination of the exciton binding energy in a quantum well is complicated by the interplay of several effects, e.g. valence band mixing, Coulomb coupling between excitons belonging to different subbands, and nonparabolicity of the bulk conduction band [34].

The oscillator strength of a QW exciton is proportional to the area S of the sample. Hence the oscillator strength per unit area is introduced:

$$\frac{f}{S} = g \frac{2}{m_0 \hbar \omega} |\langle u_{\mathrm{v}} | \hat{\boldsymbol{e}} \cdot \boldsymbol{p} | u_{\mathrm{c}} \rangle|^2 \frac{2}{\pi a_{\mathrm{B}}^2} \left| \int c(z) \, v(z) \, \mathrm{d}z \right|^2 . \qquad (1.41)$$

As the quantum well thickness decreases, f/S increases because the exciton radius a_{B} decreases.

In a QW, the lack of full translational invariance relaxes the requirement of momentum conservation in the growth direction, $\hat{\boldsymbol{e}}_z$. In contrast to bulk excitons, a QW exciton with transverse (in-plane) momentum $\hbar \boldsymbol{k}_{\|}$ interacts with a continuum of photon modes with the same transverse momentum $\hbar \boldsymbol{k}_{\|}$ but arbitrary longitudinal momentum $\hbar k_z$. This gives rise to an unassisted radiative decay. From the simplest viewpoint, the radiation emitted by the QW exciton has an escape route from the crystal. From a somewhat more sophisticated viewpoint, the oscillator describing the exciton is coupled to a continuum of oscillators describing the optical modes, which acts on the exciton as a dissipative bath and leads to an irreversible decay. Only excitons lying within the light cone, i.e. $k_{\|} \leq E_{\mathrm{ex}}/\hbar c$, can decay radiatively [35, 36]. Energy–momentum conservation prohibits radiative decay of excitons with $k_{\|} > E_{\mathrm{ex}}/\hbar c$.

The radiative decay rate of a free exciton in an isolated QW can be calculated by Fermi's golden rule under the assumption that the transverse momentum $\hbar \boldsymbol{k}_{\|}$ is conserved [37–40]. The initial state consists of an exciton with polarization vector $\hat{\boldsymbol{e}}$ and no photons present, while the final state is the crystal ground state plus a photon with a wavevector $\boldsymbol{k} = (\boldsymbol{k}_{\|}, k_z)$ and polarization σ. The squared matrix element summed over the photon polarization is calculated to be

$$\sum_{\sigma} |\langle i | H_{\mathrm{I}} | f \rangle|^2 = \frac{\pi e^2 \hbar^2}{n^2 m_0 V} \sum_{\sigma} f_{\hat{\boldsymbol{e}}} |\hat{\boldsymbol{e}} \cdot \hat{\boldsymbol{e}}_{\sigma}|^2 . \qquad (1.42)$$

Fermi's golden rule gives

$$\Gamma(k_x) = \frac{2\pi^2 e^2 \hbar^2}{n^2 m_0 V} \sum_{k_z} \sum_{\sigma} f_{\hat{e}} |\hat{e} \cdot \hat{e}_\sigma|^2 \delta \left(\hbar \omega_k - \frac{\hbar c}{n} \sqrt{k_x^2 + k_z^2} \right) . \tag{1.43}$$

Evaluating the one-dimensional density of states gives the decay rate in terms of the excitonic oscillator strength per unit area as

$$\Gamma(k_x) = \frac{2\pi^2 e^2}{n^2 m_0 c} \frac{k_0}{k_z} \sum_{\sigma} \frac{f_{\hat{e}}}{S} |\hat{e} \cdot \hat{e}_\sigma|^2 \theta(k_0 - k_x) , \tag{1.44}$$

where $k_z = \sqrt{k_0^2 - k_x^2}$.

For a given transverse wavevector $\boldsymbol{k}_{\parallel} = k_x \hat{\boldsymbol{e}}_x$, two orthogonal photon polarization vectors can be chosen as follows:

$$\hat{\boldsymbol{e}}_1 = \hat{\boldsymbol{e}}_y ,$$
$$\hat{\boldsymbol{e}}_2 = \frac{k_z \hat{\boldsymbol{e}}_x - k_x \hat{\boldsymbol{e}}_z}{k_0} . \tag{1.45}$$

For a T exciton, whose polarization lies along the y axis ($\hat{e} \| \hat{\boldsymbol{e}}_y$), $\sum_\sigma f_{\hat{e}} |\hat{e} \cdot \hat{\boldsymbol{e}}_\sigma|^2 = f_{xy}$, where f_{xy} is the oscillator strength for transverse polarization. For an L exciton, whose polarization lies along the x axis ($\hat{e} \| \hat{\boldsymbol{e}}_x$), $\sum_\sigma f_{\hat{e}} |\hat{e} \cdot \hat{\boldsymbol{e}}_\sigma|^2 = f_{xy}(k_z/k_0)^2$. For a Z exciton, whose polarization lies along the z axis ($\hat{e} \| \hat{\boldsymbol{e}}_z$), $\sum_\sigma f_{\hat{e}} |\hat{e} \cdot \hat{\boldsymbol{e}}_\sigma|^2 = f_z(k_x/k_0)^2$. Thus the radiative decay rates of T, L, and Z excitons for $k_z < k_0$ are

$$\Gamma_{\mathrm{T}}(k_x) = \frac{2\pi^2 e^2}{n^2 m_0 c} \frac{f_{xy}}{S} \frac{k_0}{k_z} , \tag{1.46}$$

$$\Gamma_{\mathrm{L}}(k_x) = \frac{2\pi^2 e^2}{n^2 m_0 c} \frac{f_{xy}}{S} \frac{k_z}{k_0} , \tag{1.47}$$

$$\Gamma_{\mathrm{Z}}(k_x) = \frac{2\pi^2 e^2}{n^2 m_0 c} \frac{f_{xy}}{S} \frac{k_x^2}{k_0 k_z} . \tag{1.48}$$

The decay rate vanishes for $k_z > k_0$.

At $\boldsymbol{k}_{\parallel} = 0$, the L and T modes have the same decay rate, $2\Gamma_0 = [2\pi e^2 / (n m_0 c)](f_{xy}/S)$. For the HH1–CB1 exciton in a 10 nm GaAs/AlGaAs quantum well, the oscillator strength $f_{xy}/S = 10 \times 10^{-5} \,\text{Å}^{-2}$, and $\hbar \Gamma_0 = 0.026$ meV. Hence the exciton radiative lifetime is $\tau_0 = 1/(2\Gamma_0) = 12$ ps. Experimental evidence of such a short radiative lifetime of free excitons in GaAs QWs has been reported [41, 42].

The above calculation of the intrinsic radiative decay of free excitons is based on the assumption of transverse-momentum conservation. Thereby, it neglects the effects of interface roughness and acoustic-phonon scattering. In real samples, the picture is not so simple; the static disorder, exciton–phonon scattering, and exciton–exciton scattering mix various $\boldsymbol{k}_{\parallel}$ states together, thus disturbing both the spatial and the temporal coherence of the excitons [43].

1.3 Semiconductor Microcavities

1.3.1 Microcavity Geometries

The density of optical modes as a function of wavelength in a microcavity depends on the dimensionality of the cavity in a manner similar to that of the density of states for electrical carriers in a quantum well, wire or dot [44]. In the bulk, three-dimensional (3D) case ($L \gg \lambda/2n$), the optical mode density increases quadratically with frequency, i.e.

$$\rho(\omega) = \frac{1}{2\pi^2 \, c^3} \, \omega^2 \,, \tag{1.49}$$

as shown in Fig. 1.3a.

In a planar cavity, as shown in Fig. 1.3b, the optical modes in the z direction (perpendicular to the mirror planes) are quantized, i.e.

$$k_z = \frac{2\pi}{d} \, m_z \,, \tag{1.50}$$

where d is the distance between the two planar mirrors, and m_z is an integer. In the transverse direction (parallel to the mirror planes), there are is a continuum of optical modes, owing to the lack of optical confinement. Hence the 2D density of states is

$$\rho(\omega) = N_z \frac{1}{2\pi \, c^2} \, \omega \,, \tag{1.51}$$

where N_z is an integer less than $2d/\lambda$, equal to the number of quantized longitudinal modes in the cavity. As shown in Fig. 1.3b, the optical-mode density increases by a step each time another half-wavelength fits between the planar mirrors (i.e. N_z jumps by one). Between these steps, the mode density increases linearly owing to modes propagating in the transverse direction.

In the zero-dimensional limit, optical modes are quantized in every direction owing to three-dimensional confinement. Therefore the density of states is a set of widely spaced δ functions, as shown in Fig. 1.3c.

Experimentally, a number of microcavity geometries have been realized in a variety of condensed-matter systems. The most widely studied semiconductor microcavity is the planar distributed Bragg reflector (DBR) cavity.

1.3.2 Bragg Mirrors

Figure 1.4 is a scanning electron microscope (SEM) picture of a planar DBR microcavity grown by molecular-beam epitaxy (MBE). The cavity is formed by two Bragg mirrors, which consist of thin layers of dielectrics with alternating low and high refractive indices. The highest reflectivity of a Bragg mirror is achieved when the layer thickness is equal to one-quarter of the optical wavelength. In this case, all Fresnel reflections add in phase and the transmitivity of the mirror decreases approximately exponentially as a function

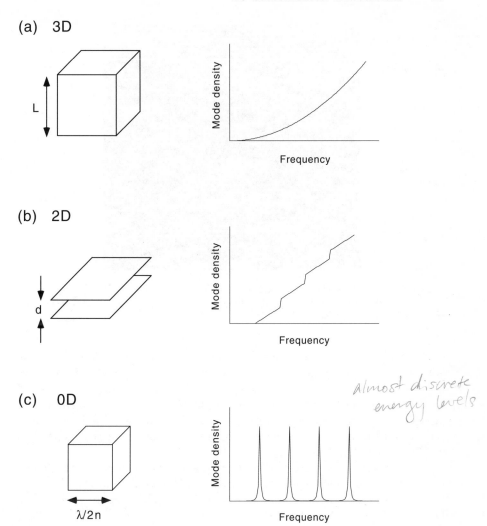

Fig. 1.3. Mode densities as a function of frequency for three optical-cavity dimensionalities. The bulk cavity in (**a**) has a dimension $L \gg \lambda/n$. The two-dimensional cavity shown in (**b**) has no modes below its cutoff frequency. In the zero-dimensional case shown in (**c**) there are well-separated, discrete modes

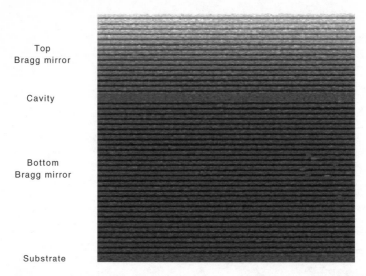

Fig. 1.4. An SEM picture of a $\lambda/2$ planar DBR microcavity. The Bragg mirrors consist of alternating AlAs and $Al_{0.15}Ga_{0.85}As$ layers. The cavity spacer layer is $Al_{0.3}Ga_{0.7}As$. The substrate is GaAs

of the mirror thickness (or the number of alternating dielectric layers). The advantages of Bragg mirrors compared with metal mirrors are that they have little loss, the reflectivity can be chosen arbitrarily (by varying the number of dielectric layers or the reflective indices of the layers), they can be epitaxially grown on semiconductor substrates, and, if made from a semiconductor material, the mirrors can be either electrically insulating or conductive, depending on the doping of the mirror layers.

The reflection phase for a wave impinging on the mirror from the spacer is zero at an interface where the refractive index goes from a higher to a lower value, and π at an interface where the refractive index goes from a lower to a higher value. Hence, by putting two mirrors together face to face, a resonant cavity is formed, with standing-wave patterns as schematically outlined in Fig. 1.5. If the mirrors are identical, the final cavity is formed in the center of a quarter-wavelength-displaced grating. (The spacer has an optical thickness of $\lambda_{Bragg}/4$.) In the cavity QED community, this cavity has been called a "one-dimensional photonic bandgap" cavity, drawing an analogy with the three-dimensional photonic bandgap structures proposed by Yablonovitch [45].

Since air has a substantially lower index of refraction than the semiconductor substrate, upon leaving the cavity on the "air side", the field sees an additional (relatively large) reflection at the semiconductor–air interface. Usually the "air side" Bragg mirror stack is designed so that this large reflection adds in phase with the other reflections, and thus the "air side" mirror has a higher reflectivity than the "substrate side" mirror, if the two

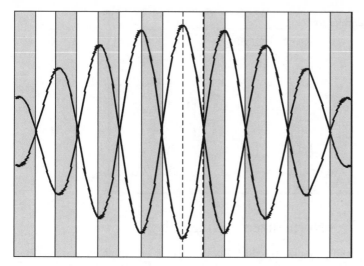

Fig. 1.5. Schematic drawing of the standing-wave patterns in a $\lambda/2$ cavity

Bragg mirrors are identical otherwise. Often this asymmetry is compensated by making the "substrate side" mirror thicker to achieve equal reflectivities as seen from the spacer looking outwards.

Figure 1.6 shows the reflectivity from a single Bragg mirror versus the wavelength [46]. The Bragg mirror has been assumed to have twenty $\lambda/4$ dielectric layers (ten periods), with alternating refractive indices of 3.0 and 3.6 (corresponding roughly to the refractive indices of a lattice-matched mirror made of $AlAs/Al_{0.15}Ga_{0.85}As$ on a GaAs substrate). The mirror has been assumed to be sandwiched between substrate materials of refractive index 3.6. As can be seen, the mirror has high reflectivity in only a certain wavelength region (often called the stopband) around the Bragg wavelength. The Bragg wavelength λ_{Bragg} is defined by the fundamental periodicity of the layers:

$$\lambda_{Bragg} = 2(l_1 n_1 + l_2 n_2) \,, \tag{1.52}$$

where l_i and n_i are the thickness and refractive index of the layers $i = 1, 2$. The width of the stopband is approximately given by [47]

$$\Delta\lambda_{stopband} = \frac{2\lambda_{Bragg}\,\Delta n}{\pi\,n_{eff}} \,, \tag{1.53}$$

where $\Delta n = |n_2 - n_1|$, and n_{eff} is the effective index of the mirror. For small refractive-index differences n_{eff} can be replaced by the arithmetic mean of the refractive indices in the stack, while for large differences the geometric mean is more suitable.

In Fig. 1.7, the reflection phase of the Bragg mirror (modulo 2π) is plotted as a function of the wavelength. It can be seen from Fig. 1.7 that the reflection phase is π only exactly at the Bragg wavelength. In the stopband the reflection

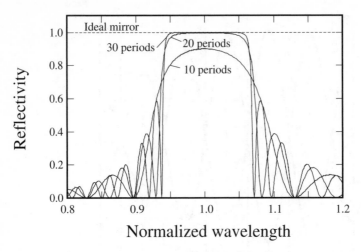

Fig. 1.6. Computed plane wave reflectivity of a Bragg mirror at normal incidence as a function of the normalized wavelength $\lambda/\lambda_{\text{Bragg}}$. The refractive indices have been assumed to be 3.6 and 3.0. Curves have been drawn for mirrors of 10 periods (20 layers), 20 periods, and 30 periods

phase dispersion (the slope of the curve) is small. This is compensated by high dispersion at each edge of the stopband. The dispersion in the center of the stopband is independent of the number of layers in the mirror; it depends only on the relative refractive-index difference of the mirror. This effect can be attributed to the fact that the reflection occurs gradually as the wave

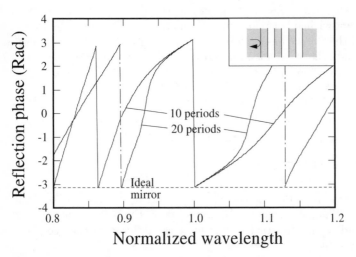

Fig. 1.7. Computed reflection phase of a Bragg mirror as a function of the normalized wavelength $\lambda/\lambda_{\text{Bragg}}$. The reference location is shown in the inset, where the *shaded areas* represent the $n = 3.6$ layers

penetrates the Bragg stack. Hence the Bragg mirror can, in the stopband, be modeled as an ideal mirror with fixed reflection phase, sitting some distance l_{pen} away from the reference plane. The reflected wave will have a phase delay of $\exp(-\mathrm{i}\,2k\,l_{\text{pen}})$ compared with the incident wave. The penetration length of the wave into the Bragg mirror at $\lambda = \lambda_{\text{Bragg}}$ is roughly

$$l_{\text{pen}} \approx \frac{3}{2}\frac{\lambda_{\text{Bragg}}}{n_{\text{eff}}} . \tag{1.54}$$

Hence, even for an AlAs/Al$_{0.15}$Ga$_{0.85}$As mirror with a relatively high refractive-index ratio, the penetration depth into the mirror is about 1.5 wavelengths. Hence the shortest microcavity in this material that can be constructed with epitaxial Bragg mirrors is $3\lambda_{\text{Bragg}}$ long.

The reason the stopband is limited in a Bragg mirror is that this mirror is a resonant structure. The interface reflections all add in phase only at λ_{Bragg}. At longer or shorter wavelengths (shorter or longer $\boldsymbol{k}$ vectors) the reflections start to add out of phase and the total reflectivity goes down. For a planar mirror, to a first-order approximation, only the wavevector component k_z is important for the resonance condition. Therefore, when keeping the wavelength constant at the Bragg wavelength but changing the angle of incidence from the normal direction to higher angles, there is a finite angular range where the reflectivity remains high. The stopband angle can be estimated from (1.53) to be

$$\Delta\theta_{\text{stopband}} = \pm\sqrt{\frac{2\Delta\lambda_{\text{stopband}}}{\lambda_{\text{Bragg}}}} = \pm\sqrt{\frac{\Delta n}{\pi n_{\text{eff}}}} . \tag{1.55}$$

Putting in the numbers for an AlAs/Al$_{0.15}$Ga$_{0.85}$As Bragg mirror, the mirror is only highly reflecting for angles smaller than 0.48 radians, or 28 degrees. In Fig. 1.8, the plane wave reflectance of this Bragg mirror at the Bragg wavelength is plotted as a function of the incidence angle in the GaAs material. It is seen that the P-polarized wave has a slightly smaller stopband than the S-polarized wave. Near the Brewster angle $\theta_{\text{Brewster}} = \arctan(n_2/n_1) \sim 40°$, the P wave sees a totally transparent mirror, whereas the S wave sees an oscillating reflectance with finite amplitude. At angles greater than $\arcsin(n_2/n_1) \sim 50°$, the incident waves are evanescent in the AlAs ($n = 3.0$) layers. Nonetheless, the reflectivity approaches unity (total internal reflection) only for angles greater than $\arcsin(n_2/n_{\text{eff}}) \sim 65°$. For incidence angles between $50°$ and $65°$ degrees the field tunnels evanescently through the Bragg stack. The angular region between $28°$ and $65°$ is sometimes referred to as the "open window" of the Bragg mirror.

1.3.3 Planar DBR Microcavities

Since a Bragg mirror is transparent outside the wavelength and angular stopband, in an one-dimensional microcavity constructed from two planar DBR

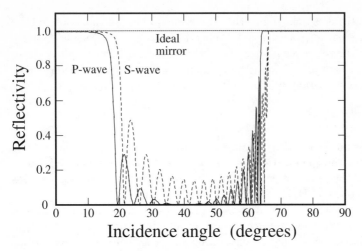

Fig. 1.8. Computed plane wave reflectivity of a Bragg mirror at the Bragg wavelength as a function of the incidence angle

mirrors there exist not only the Bragg resonant modes within the stopband, but also leaky modes inside the open window of the Bragg mirrors, and guided modes originating from total internal reflection [48]. Figure 1.9 shows the distribution of the longitudinal-mode density in a planar DBR microcavity. For S-polarization, the lowest-order Bragg resonant mode, indicated by A, satisfies the resonant condition

$$k_z = k \cos(\theta) = \frac{\pi}{n \, L_c} \, , \tag{1.56}$$

where L_c is the cavity length, and n is the refractive index. For $k > \pi/(n \, L_c)$, the resonance condition can be fulfilled at an angle

$$\theta_k = \arccos\left(\frac{\pi}{k \, n \, L_c}\right) \tag{1.57}$$

Around the Bragg resonant mode, there are spaces of small mode density which correspond to the stopband of the DBR. The halftone area around the stopband corresponds to weakly resonant leaky waves. The boundary, indicated by B, shows the critical angle for total internal reflection. Below the boundary, there are strongly resonant modes propagating in the transverse direction, called "guided modes". The dispersion of the fundamental propagation mode B is shown by C. The number of guided modes increases as the cavity length decreases. For P polarization, the Brewster angle breaks the curve of the Bragg resonant mode off, as indicated by D in Fig. 1.9b.

In an ideal planar microcavity, where the mirror reflectivity is equal to unity, a Bragg resonant mode corresponds to a plane wave propagating perpendicular to the cavity. However, in a real microcavity, the nonunity mirror reflectivity induces a finite cavity dissipation, and leads to a cavity quasi-

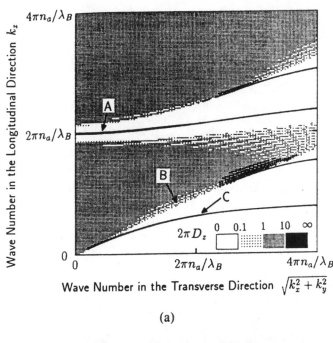

(a)

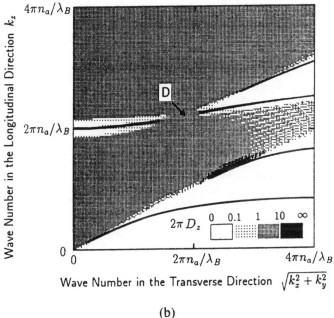

(b)

Fig. 1.9. Distribution of longitudinal-mode density in a λ cavity: (**a**) S polarization, (**b**) P polarization

mode with a finite angular spread. The full width at half maximum (FWHM) of the angular spread of the cavity quasi-mode is [46]

$$\Delta\theta_{\mathrm{FWHM}} = \sqrt{\frac{2\lambda[1 - (R_1 R_2)^{1/2}]}{\pi L (R_1 R_2)^{1/4}}} \,, \tag{1.58}$$

where R_1 and R_2 are the reflectivities of the two mirrors. Correspondingly, the cavity quasi-mode has a finite effective radius r_{m}:

$$r_{\mathrm{m}} = \sqrt{\frac{\lambda L (R_1 R_2)^{1/4}}{\pi [1 - (R_1 R_2)^{1/2}]}} \,. \tag{1.59}$$

As the mirror reflectivity increases, the mode is spread over a larger volume, and its angular spread becomes narrower.

The photon energy dispersion in a planar cavity can be described as

$$\omega_{\mathrm{ph}}(k_\parallel) = \frac{\hbar c}{n_{\mathrm{cav}}} \sqrt{k_\parallel^2 + k_z^2} \,, \tag{1.60}$$

where $k_\parallel$ is the photon $\boldsymbol{k}$ vector parallel to the mirror layers, and its value can be continuously varied. On the other hand, $k_z = m\pi/L$ is quantized. In the regime where $k_\parallel \ll k_z$, (1.60) can be approximated as

$$\begin{aligned} \omega_{\mathrm{ph}}(k_\parallel) &\simeq \frac{\hbar c}{n_{\mathrm{cav}}} k_z + \frac{\hbar c}{2 n_{\mathrm{cav}} k_z} k_\parallel^2 \\ &= \frac{\hbar c}{n_{\mathrm{cav}}} k_z + \frac{\hbar^2 k_\parallel^2}{2 m_{\mathrm{ph}}} \,. \end{aligned} \tag{1.61}$$

This indicates the cavity photon has an effective mass

$$m_{ph} = \frac{n_{\mathrm{cav}} k_z}{\hbar c} \,. \tag{1.62}$$

1.3.4 Exciton–Photon Coupling

Consider a GaAs quantum well (QW) located inside a planar DBR microcavity, as shown in Fig. 1.4. The center-of-mass wavefunction of the two-dimensional (2D) excitons confined in the QW can be written as

$$F_{k_\parallel, n}(\boldsymbol{r}_\parallel, z) = \sqrt{\frac{2}{L_{\mathrm{w}}}} \sin\left(\frac{\pi n (z - z_0)}{L_{\mathrm{w}}}\right) \frac{e^{i k_\parallel \cdot \boldsymbol{r}_\parallel}}{\sqrt{S}} \,, \tag{1.63}$$

where L_{w} is the QW thickness, z_0 indicates the position of the QW inside the cavity, and n is an integer number.

A longitudinal cavity photon mode of the cavity can be described by

$$v_{k'_\parallel, n'}(\boldsymbol{r}_\parallel, z) = \sqrt{\frac{2}{L_{\mathrm{w}}}} \sin\left(\frac{\pi n' z}{L_{\mathrm{eff}}}\right) \frac{e^{i k'_\parallel \cdot \boldsymbol{r}_\parallel}}{\sqrt{S}} \,, \tag{1.64}$$

where L_{eff} is the effective cavity length.

The coupling strength of the QW exciton to the cavity photon mode is embodied in the linear coupling constant Ω:

$$\Omega = e\sqrt{\frac{f}{8\epsilon L_{\mathrm{w}} m_{\mathrm{e}} S}} \int \mathrm{d}^2 r_{\parallel} \, \mathrm{d}z \, F_{k_{\parallel},n}(r_{\parallel}, z) \, v_{k'_{\parallel},n'}(r_{\parallel}, z) \,, \tag{1.65}$$

where f/S is the excitonic oscillator strength per unit area.

From (1.65), if the QW is located at a node of the intracavity field, Ω is close to zero, and thus the QW excitons are decoupled from the cavity photon mode. However, if the QW is located at an antinode of the intracavity field, the exciton–photon coupling is significantly enhanced .

Meanwhile, both QW excitons and cavity photons couple to the outside reservoirs; for example, the QW excitons can decay by emitting photons in the transverse direction, and the cavity photons can leak out of the cavity. Depending on the relative magnitude of the exciton–photon coupling constant Ω compared with the exciton and photon decay rates γ, γ_{c}, there are two regimes. If $\Omega \ll \gamma, \gamma_{\mathrm{c}}$, the system is in the weak-coupling regime. In the weak-coupling regime, the spontaneous-emission rate and pattern can be significantly modified by a cavity [49]. Nonetheless, the spontaneous-emission remains an irreversible process. However, if $\Omega \gg \gamma, \gamma_{\mathrm{c}}$, the system is in the strong-coupling regime, where the spontaneous emission becomes reversible [50].

2. Microcavity Exciton Polaritons

2.1 Strong Coupling – Theory

As semiconductor fabrication technology developed, it became possible to fabricate high-Q semiconductor microcavities. Thus the study of semiconductor cavity QED started in the weak-coupling regime [49] and then entered the strong-coupling regime, where the exciton–photon coupling constant becomes larger than the exciton and cavity photon decay rates [50]. In a high-Q semiconductor microcavity, the strong exciton–photon coupling leads to the formation of two new eigenstates of the exciton–photon coupled system, called "microcavity exciton polariton" states. The energy separation between the two polariton states increases as the exciton–photon coupling increases. This exciton–polariton normal-mode splitting is the solid-state analog of the vacuum Rabi splitting in the atom–cavity case. In the time domain, the strong exciton–photon coupling makes the spontaneous-emission process reversible; namely, the emission from the microcavity shows an oscillation, instead of the usual exponential decay. This is because photons emitted by excitons are reabsorbed and reemitted a number of times before exiting the cavity. Hence the excitation energy of the system is transferred back and forth between the QW exciton state and the cavity photon state, leading to a Rabi oscillation. This process can be modeled as a coupled pendulum system where the two pendulums correspond to the microcavity mode field at the frequency ω_c and the exciton at the frequency ω_{ex}. The QW exciton state and the resonant-cavity photon state form a simple system of two harmonic oscillators coupled through the light–matter interaction.

Different theoretical schemes have been proposed to describe the exciton–photon coupled system in a planar DBR microcavity. They range from a semiclassical linear-dispersion model, in which the active medium is represented by a Lorentz oscillator [50,51], to a full quantum theory of the exciton–photon interaction in which the complex mixed-mode energies are derived from the poles of the Green function [52]. The semiclassical theory is convenient for the computation of reflectivity, transmission, and absorption. The quantum theory, on the other hand, is more appropriate for the computation of photoluminescence (PL).

2.1.1 Semiclassical Theory

In the semiclassical theory of microcavity exciton polaritons, Maxwell's equations are solved together with a constitutive relation between the electric field and displacement field. In calculating the optical response of a semiconductor microcavity, each layer except the active medium (the QWs) can be modeled by a local, frequency-independent dielectric constant. The QWs contain in addition the excitonic contribution, which depends on frequency with a resonant form and gives an intrinsically nonlocal dielectric response [53]. We take the z axis along the growth direction, and assume the in-plane wavevector $\boldsymbol{k}_{\parallel} = k\,\boldsymbol{e}_x$ to lie along the x axis. The constitutive relation for $\boldsymbol{E}$ and $\boldsymbol{D}$ within each QW is taken to be of the form

$$\boldsymbol{D}(z) = \epsilon_\infty\,\boldsymbol{E}(z) + 4\pi \int \chi(\omega, k; z, z')\,\boldsymbol{E}(z')\,\mathrm{d}z'\,. \tag{2.1}$$

The nonlocal response function $\chi(\omega, k; z, z')$ associated with the excitonic resonance is given in linear-response theory as

$$\chi(\omega, k; z, z') = \frac{g_{\rm s}|\langle u_{\rm c}|e\boldsymbol{r}|u_{\rm v}\rangle|^2}{\hbar[\omega_{\rm ex}(k_x) - \omega - i\gamma]}\,\rho(z)\,\rho(z')\,, \tag{2.2}$$

where $\rho(z) = F_{\rm 2D}(\rho = 0)\,c(z)\,v(z)$ is the exciton envelope function at zero electron–hole separation, $\omega_{\rm ex}(k)$ is the exciton energy including spatial dispersion, $g_{\rm s}$ is a factor of order unity which accounts for the presence of spin–orbit interaction, and γ is a broadening term which describes nonradiative decay processes of the exciton.

Maxwell's equations, with the above constitutive relation and the appropriate boundary conditions at each interface, completely specify the electrodynamic problem to be solved. Maxwell's equations decouple into TE (or S) and TM (or P) polarizations in this case. The electric field is polarized along the y axis for TE polarization, and along the x and z axes for TM polarization. The z polarization is not coupled to the heavy-hole exciton. For a layered system, it is useful to formulate Maxwell's equations in a transfer-matrix scheme [54]. The transfer matrix M is a 2×2 matrix, which acts on a basis of right-traveling and left-traveling waves. It propagates the components of the electric field from a point $z_{\rm l}$ to a point $z_{\rm r}$ in the structure. It is a unimodular matrix when the refractive indexes of the media on the left and right are the same; otherwise, $\det(M) = n_{\rm l}/n_{\rm r}$. The reflection and transmission coefficients for a system characterized by a transfer matrix M are given by

$$r = -\frac{M_{21}}{M_{22}}\,, \qquad t = \frac{\det(M)}{M_{22}}\,. \tag{2.3}$$

The reflectivity and transmittivity are

$$R = |r|^2\,, \qquad T = \frac{n_{\rm r}}{n_l}|t|^2 = \frac{n_r}{n_l}\frac{1}{|M_{22}|^2}\,, \tag{2.4}$$

and the absorption is $A = 1 - R - T$. The poles of the reflection and transmission coefficients, namely, the solutions of the equation $M_{22} = 0$ as a function of frequency and in-plane wavevector, give the complex energies of resonance. The real parts of the complex energies correspond to the energies of the microcavity exciton polaritons, and the imaginary parts give the decay rates.

When the nonradiative broadening γ is sufficiently large, a local approximation can be adopted, where the exciton contribution can be included in the dispersive dielectric constant of a Lorentz oscillator [55]:

$$\epsilon(E) = \epsilon_\infty + \frac{e^2 h^2}{m_0 \epsilon_0 L_w} \frac{f}{S} \frac{1}{\hbar^2(\omega_{ex}^2 - \omega^2 - i\gamma\omega)}, \tag{2.5}$$

where f/S is the excitonic oscillator strength per unit area, and L_w is the quantum well thickness.

From the transfer matrix of the whole microcavity structure, the transmission T, reflectivity R, and absorption A can be calculated numerically. In the high-reflectivity limit $1 - R \ll 1$ ($R_1 = R_2 = R$), at resonance, $\omega_c = \omega_{ex}$, the following analytical formulas for the splittings in T, R, and A can be derived [56]:

$$\Delta\omega_T = 2\sqrt{\sqrt{V^4 + 2V^2\gamma(\gamma + \gamma_c)} - \gamma^2}, \tag{2.6}$$

$$\Delta\omega_R = 2\sqrt{\sqrt{V^4\left(1 + \frac{2\gamma}{\gamma_c}\right)^2 + 2V^2\gamma^2\left(1 + \frac{\gamma}{\gamma_c}\right)} - 2V^2\frac{\gamma}{\gamma_c} - \gamma^2}, \tag{2.7}$$

$$\Delta\omega_A = 2\sqrt{V^2 - \frac{1}{2}(\gamma^2 + \gamma_c^2)}, , \tag{2.8}$$

where

$$V = \sqrt{\frac{2c\Gamma_0}{n_{cav} L_{eff}}}, \tag{2.9}$$

$$\gamma_c = \frac{c(1 - R)}{2n_{cav} L_{eff}}, \tag{2.10}$$

$L_{eff} = L_c + L_{DBR}$ is the effective cavity length, which includes the penetration length into the DBR mirrors L_{DBR}, n_{cav} is the effective refractive index of the cavity, and $\Gamma_0 = e^2/(4\epsilon_0 n_{cav} m_0 c)(f/S)$ is the decay rate of excitons with $k = 0$ in a single QW.

As compared with the vacuum Rabi splitting in an atomic cavity, we can see the following differences. (i) The numerator in the square root of the expression for V contains an additional factor of two owing to the multiple reflections from the QW. (ii) The additional cavity length L_{DBR}, originating from the energy dependence of the DBR phase, is often much larger than the cavity length L_c, and thus leads to a reduction of the Rabi splitting. (iii) The splittings in transmission, reflectivity, absorption, and the zeros of

M_{22} coincide only when the nonradiative broadening $\gamma = 0$. When $\gamma \neq 0$, the condition for a splitting in the absorption is more stringent than that for the reflectivity or transmission, and the splitting in the absorption is smaller than the splitting in the reflectivity and transmission.

Note that the microcavity polariton splitting is reduced compared with the bulk polariton splitting by a factor of the order of $\sqrt{L_w/L_{\mathrm{eff}}}$. This is because QW excitons interact with the radiation field in only a region of thickness L_w along the growth direction, and the exciton and photon wavefunctions do not completely overlap as they do in the bulk.

2.1.2 Quantum Theory

The quantum theory of a microcavity exciton–photon coupled system is based on a Hamiltonian obtained through a microscopic theory of the coupling between the QW exciton and the quantized electromagnetic modes of the surrounding medium. The normal modes which diagonalize the Hamiltonian are the polariton modes. However, in the quantum theory, it is not easy to give a realistic description of the DBRs, in particular, to describe the frequency dependence of the phase of the DBR reflection coefficient.

Consider a QW embedded in a planar DBR microcavity. The microscopic Hamiltonian of a Wannier exciton coupled to the radiation field is given by [57]

$$
\begin{aligned}
H = &\sum_{\boldsymbol{k}_\parallel} \hbar\omega_{\mathrm{ex}}(\boldsymbol{k}_\parallel) b^\dagger_{\boldsymbol{k}_\parallel} b_{\boldsymbol{k}_\parallel} + \sum_{\boldsymbol{k}_\parallel} \hbar\omega_{\mathrm{ph}}(\boldsymbol{k}_\parallel) b^\dagger_{\boldsymbol{k}_\parallel} b_{\boldsymbol{k}_\parallel} \\
&+ \sum_{\boldsymbol{k}_\parallel} iC_k (a^\dagger_{\boldsymbol{k}_\parallel} + a_{-\boldsymbol{k}_\parallel})(b_{\boldsymbol{k}_\parallel} - b^\dagger_{-\boldsymbol{k}_\parallel}) \\
&+ \sum_{\boldsymbol{k}_\parallel} D_{\boldsymbol{k}_\parallel}(a^\dagger_{\boldsymbol{k}_\parallel} + a_{-\boldsymbol{k}_\parallel})(a_{\boldsymbol{k}_\parallel} + a^\dagger_{-\boldsymbol{k}_\parallel}) \,,
\end{aligned}
\tag{2.11}
$$

where $a_{\boldsymbol{k}_\parallel}$ and $b_{\boldsymbol{k}_\parallel}$ are the lowering operators for QW exciton and cavity photon modes with transverse momentum $\hbar\boldsymbol{k}_\parallel$. Owing to the in-plane translational invariance of the electronic and photonic system, the transverse momentum is conserved, namely, a QW exciton state with a well-defined transverse momentum $\hbar\boldsymbol{k}_\parallel$ is coupled to a single cavity photon mode with the same transverse momentum $\hbar\boldsymbol{k}_\parallel$. In the linear regime, different $\boldsymbol{k}_\parallel$ modes are decoupled. This allows us to consider a single cavity mode and a single QW exciton mode with the same $\boldsymbol{k}_\parallel$. In the following, we set:

$$
a_{\boldsymbol{k}_\parallel} = a_{-\boldsymbol{k}_\parallel} \,, \quad b_{\boldsymbol{k}_\parallel} = b_{-\boldsymbol{k}_\parallel} \,,
\tag{2.12}
$$

and omit the subscript $\boldsymbol{k}_\parallel$ from now on. The Hamiltonian is then

$$
H = \hbar\omega_{\mathrm{ex}} b^\dagger b + \hbar\omega_c a^\dagger a + iC(a + a^\dagger)(b - b^\dagger) + D(a^\dagger + a)(a + a^\dagger) \,.
\tag{2.13}
$$

The fourth term corresponds to the photon self-interaction term $\boldsymbol{A}^2$. It can be neglected when the dispersion is taken sufficiently close to the exciton resonance and the cavity width is narrow enough [59]. Setting $C = \hbar\Omega$, where Ω represents the exciton–photon coupling constant, and using the rotating-wave approximation, we obtain

$$H = \hbar\omega_{\text{ex}}b^\dagger b + \hbar\omega_{\text{ph}}a^\dagger a + i\hbar\Omega(a^\dagger b - ab^\dagger)$$
$$+ \sum_k \hbar g_{\text{ex},k}(b_k^\dagger b + b^\dagger b_k) + \sum_{k'} \hbar g_{\text{ph},k'}(a_{k'}^\dagger a + a^\dagger a_{k'}) . \tag{2.14}$$

In the above equation, we have added coupling terms to exciton and photon reservoirs with coupling constants $g_{\text{ex},k}$ and $g_{\text{ph},k'}$. In a real cavity, the exciton and photon modes are coupled to a continuum of modes, which leads to dissipation. The coupling can be phonon scattering in the case of the exciton, or cavity damping in the case of the photon. Introducing the Markov approximation for the damping terms, we can derive the Heisenberg equations of motion for the photon and exciton field operators a and b, respectively [58]:

$$\frac{da}{dt} = -i\omega_c a - \Omega b - \frac{\gamma_c}{2}a + F_a, \tag{2.15}$$

$$\frac{db}{dt} = -i\omega_{\text{ex}}b - \Omega a - \frac{\gamma}{2}b + F_b, \tag{2.16}$$

where F_a and F_b are the noise terms necessary to preserve the commutator, and are thus related to the damping terms. However, since in the strong-coupling regime the exciton–photon coupling constant is larger than the exciton and photon decay rates, the noise terms can be neglected if we are not interested in the noise properties. The solutions for above coupled equations are

$$a(t) = \frac{(\omega_c - \omega_- - i\gamma_c/2)a(0) + i\Omega b(0)}{\Delta}e^{-i\omega_+ t}$$
$$+ \frac{(-\omega_c + \omega_+ + i\gamma_c/2)a(0) - i\Omega b(0)}{\Delta}e^{-i\omega_- t} , \tag{2.17}$$

$$b(t) = \frac{(\omega_{\text{ex}} - \omega_- - i\gamma/2)b(0) - i\Omega a(0)}{\Delta}e^{-i\omega_+ t}$$
$$+ \frac{(-\omega_{\text{ex}} + \omega_+ + i\gamma/2)b(0) + i\Omega a(0)}{\Delta}e^{-i\omega_- t} , \tag{2.18}$$

where $a(0)$ and $b(0)$ are the values of a and b at $t = 0$, and

$$\omega_\pm = \frac{[\omega_{\text{ex}} + \omega_c - i(\gamma_c + \gamma)/2] \pm \sqrt{[\omega_c - \omega_{\text{ex}} - i(\gamma_c - \gamma)/2]^2 + 4\Omega^2}}{2} . \tag{2.19}$$

The splitting is

$$\Delta = \omega_+ - \omega_- = \sqrt{[\omega_c - \omega_{\text{ex}} - i(\gamma_c - \gamma)/2]^2 + 4\Omega^2} . \tag{2.20}$$

It is useful to separate the real and imaginary parts of $\omega_\pm$:

$$\omega_\pm \equiv \omega'_\pm + i\gamma_\pm$$

$$= \frac{\omega_{\mathrm{ex}} + \omega_c}{2} \pm \frac{(a^2 + b^2)^{1/4}}{2} \cos\left[\frac{1}{2}\arctan\left(\frac{b}{a}\right)\right]$$

$$-i\left\{\frac{\gamma_{\mathrm{ex}} + \gamma_{\mathrm{ph}}}{4} \mp \frac{(a^2 + b^2)^{1/4}}{2} \sin\left[\frac{1}{2}\arctan\left(\frac{b}{a}\right)\right]\right\}, \qquad (2.21)$$

with

$$a = (\omega_c - \omega_{\mathrm{ex}})^2 - \frac{(\gamma_c - \gamma)^2}{4} + 4\Omega^2, \qquad (2.22)$$

$$b = -(\gamma_c - \gamma)(\omega_c - \omega_{\mathrm{ex}}). \qquad (2.23)$$

From (2.17), the intensity of the photon field at resonance $\omega_{\mathrm{ex}} = \omega_c$ and for $a(0) = 0$ has the simple form

$$I_c(t) = \langle a^\dagger(t)a(t)\rangle$$

$$= \frac{4b(0)^2\Omega^2}{\Delta\Delta^*} e^{-(\gamma+\gamma_c)t/2} \sin\left(\frac{\Delta}{2}t\right)\sin\left(\frac{\Delta^*}{2}t\right). \qquad (2.24)$$

Therefore the microcavity emission intensity shows a temporal oscillation, i.e. a Rabi oscillation. The oscillation period is equal to $2\pi/(\Delta + \Delta^*)$.

In the frequency domain, the spectrum of the system is defined as the Fourier transform of the correlation function. In the case of an ergodic, stationary process, we have [60]

$$S(\omega) = \int_0^\infty e^{-i\omega t}\langle a^\dagger(t)a(0)\rangle dt + \mathrm{c.c.} \qquad (2.25)$$

If the cavity damping is moderate, the microcavity exciton–photon coupling is almost ergodic and stationary. From (2.25), it can be shown that the photoluminescence spectrum is a superposition of two Lorentzian lines [58]:

$$S(\omega) = \frac{A_+}{(\omega - \omega'_+)^2 + \gamma_+^2} + \frac{A_-}{(\omega - \omega'_-)^2 + \gamma_-^2}, \qquad (2.26)$$

where $A_\pm$ are slowly varying functions of ω near the peaks and can be considered as constant. Therefore the splitting in photoluminescence is

$$\Delta\omega_{\mathrm{PL}} = \sqrt{2\Delta\sqrt{\Delta^2 + \gamma_0^2} - \Delta^2 - \gamma_0^2}, \qquad (2.27)$$

where $\gamma_0 = (\gamma_c + \gamma)/2$. The polariton dispersion can also be obtained by solving for the poles of the exciton Green function [52, 56]. This gives the same results.

Figure 2.1 shows the calculated splitting in the absorption ($\Delta\omega_A$), transmission ($\Delta\omega_T$), reflection ($\Delta\omega_R$), and photoluminescence ($\Delta\omega_{\mathrm{PL}}$), and the Rabi splitting ($\mathrm{Re}(\Delta)$) as a function of the nonradiative broadening γ in a λ cavity with one QW at the center. It can be seen that the splittings are all different: they tend to be close to each other when $\gamma, \gamma_c \ll V$. But they

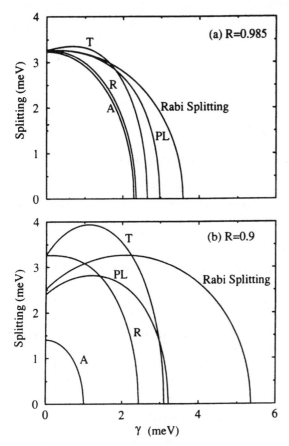

Fig. 2.1. Calculated splitting in the absorption, transmission, reflection, and photoluminescence, as well as the Rabi splitting, as a function of the nonradiative broadening γ in a λ cavity with one QW at the center

differ considerably for a lower cavity finesse (Fig. 2.1b). The smallest splitting is always in absorption, and for $\gamma_c \ll V$ the splitting in the reflectivity is close to that in the absorption. If γ is not too large, the splitting in the transmission is the largest among all the splittings. The "Rabi splitting" is not coincide with any of the other splittings.

If several QWs are embedded in a microcavity, the effect of multiple QWs can be taken into account analytically, in the transfer matrix method, by working to linear order in Γ_0. This amounts to neglecting multiple interference, or polariton effects within multiple quantum wells (MQWs). It is also equivalent to summing the squared electric field at the QW positions [59]. As a result, a MQW with N wells and a period L_s can be replaced by an effective number N_{eff} of QWs:

$$N_{\text{eff}} = \frac{N}{2} + \frac{1}{2}\frac{\sin(NkL_{\text{s}})}{\cos(kL_{\text{s}})} \,, \tag{2.28}$$

and the splitting becomes

$$\Delta_N = \Delta\sqrt{N_{\text{eff}}}\,. \tag{2.29}$$

2.2 Strong Coupling – Experiment

2.2.1 Microcavity Exciton Polariton Splitting and Oscillation

Ever since Weisbuch et al. reported the first observation of coupled exciton–photon mode splitting in a high-Q semiconductor microcavity [50], there have been extensive experimental studies of microcavity exciton polaritons in both absorption and emission measurements. As shown in Fig. 2.2, the exciton polariton dispersion curves, obtained from angle-resolved photoluminescence measurements, show anticrossing [61]. This gives a clear evidence of strong exciton–photon coupling in a semiconductor microcavity. In the time domain, microcavity emission under resonant excitation by femtosecond laser pulses shows beats corresponding to oscillation between a QW exciton state and a cavity photon state [62,63]. Interferometric pump–probe measurements by Norris et al. suggest a coherent evolution of the cavity polarization [62]. The temporal oscillation of the coherent four-wave-mixing signal from a microcavity confirms the creation of a coherent superposition state between the two

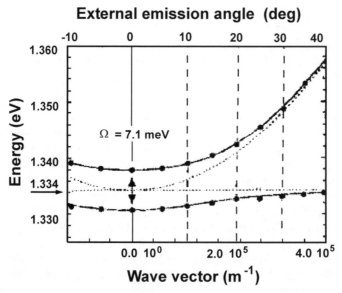

Fig. 2.2. Exciton polariton dispersion curves obtained from angle-resolved photoluminescence measurements

normal modes of the composite system by coherent short pulses. It demonstrates a coherent energy exchange between excitons and cavity photons [64].

In a GaAs/AlAs λ cavity containing a single InGaAs QW, the exciton polariton normal-mode splitting is around 4 meV. By incorporating multiple QWs in the cavity and/or increasing the optical confinement through selective oxidation of AlAs mirror layers, the exciton polariton normal-mode splitting can be enhanced to close to 10 meV. In addition, the splitting can be observed up to room temperature [65, 66]. In microcavities containing ZnCdSe QWs, the large excitonic oscillator strength results in a 17 meV exciton polariton normal-mode splitting [67].

2.2.2 Collapse and Revival of Exciton Polariton Oscillation

In InGaAs/GaAs QWs, strain induced by lattice mismatch pushes the heavy-hole (HH) and light-hole (LH) exciton states further apart in energy. But in GaAs/AlGaAs QWs, the HH and LH exciton states are close to each other in energy. The strong coupling of both the HH and the LH exciton to the cavity photon state in a high-Q microcavity results in an interesting phenomenon: collapse and revival of the Rabi oscillation [69].

The microcavity sample used in our experiment described below, was grown by MBE, and consists of a single 20 nm GaAs QW located in the middle of a $\lambda/2$ DBR cavity. The top and bottom mirrors consist of 15.5 and 30 pairs, respectively, of $Al_{0.15}Ga_{0.85}As$ and AlAs layers. The top $Al_{0.3}Ga_{0.7}As$ spacer layer is tapered along one direction of the sample so that the resonant photon frequency of the cavity varies with sample position. The coupled exciton polariton modes were probed by an absorption measurement at 4.2 K.

When the cavity photon energy is tuned to the QW HH exciton energy, the strong coupling of the QW HH exciton state to the cavity photon state results in two HH exciton polariton states, as shown in Fig. 2.3a. Similarly, when the cavity photon energy is tuned to the QW LH exciton energy, the strong coupling of the LH exciton to the resonant photon mode leads to two LH exciton polariton peaks in the absorption spectrum, as shown in Fig. 2.3b.

Figure 2.4 shows the measured exciton polariton energies obtained as we tuned the cavity resonant frequency by shifting the excitation spot on the sample. The exciton polariton dispersion curves feature two anticrossings, indicating the strong coupling of both the QW HH exciton state and the LH exciton state to the cavity photon state. The solid lines in Fig. 2.4 are theoretically fitted exciton polariton dispersion curves.

We measured the temporal evolution of the microcavity emission using an AC balanced homodyne detector. The advantages of the balanced homodyne detection scheme are high sensitivity and high temporal resolution [68].

Trace A in Fig. 2.5 was taken at the HH exciton resonance, i.e. position A in Fig. 2.4. The measured microcavity emission shows a regular oscillation due to the beating between the two HH exciton polariton states (i.e. polaritons 1 and 2 in Fig. 2.4). The contribution from the third LH exciton polariton is

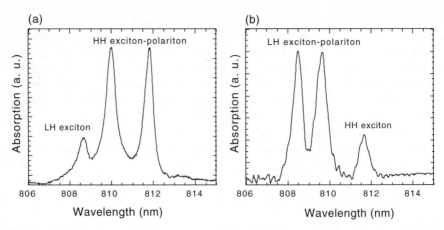

Fig. 2.3. Measured absorption spectrum at (**a**) HH exciton resonance, (**b**) LH exciton resonance

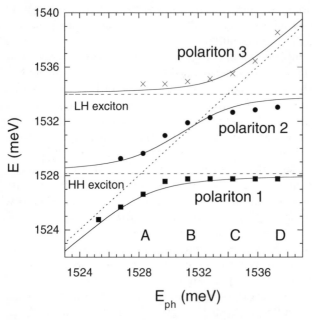

Fig. 2.4. Exciton polariton spatial dispersion curves deduced from absorption measurement. The *solid lines* are the theoretically fitted exciton polariton dispersion curves. The dashed and dotted lines are the uncoupled exciton and photon dispersion curves, respectively

negligible at this point. The first peak of trace A in Fig. 2.5 corresponds to the reflected pump pulse, which marks the zero of the time scale. After the pump pulse creates excitons in the QW, the excitons radiatively recombine

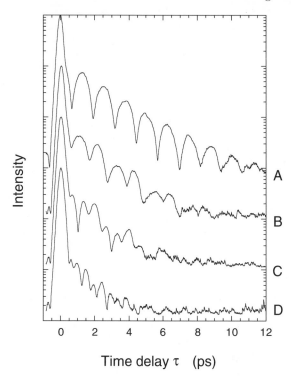

Fig. 2.5. Measured microcavity emission as a function of the time delay τ at positions A, B, C, D of Fig. 2.4.

and emit photons into the cavity mode. Some of the photons escape from the cavity and produces the first emission peak (the second peak in trace A). The rest of the photons are reflected by the cavity mirror and reabsorbed by the QW to create excitons again, leading to the valley following the first emission peak. Then the excitons emit photons again, resulting in the second emission peak. Thus the microcavity system oscillates back and forth between the QW exciton state and the cavity photon state, leading to a Rabi oscillation. The oscillation is damped owing to the cavity loss and exciton scattering. The oscillation period was measured to be about 1.2 ps, which is in good agreement with the spectral splitting of about 2.1 nm (4.1 meV) between the two HH exciton polariton states.

However, as we tuned the cavity photon energy away from the HH exciton energy, the oscillation became irregular (see traces B, C, and D in Fig. 2.5). This is because the amplitude of the LH exciton polariton state (polariton 3 in Fig. 2.4) is no longer negligible, and the beating of the three polariton modes results in a nonsinusoidal oscillation. The temporal behavior of the microcavity emission at various cavity detunings can be well predicted by our theoretical model based on three coupled harmonic oscillators [69].

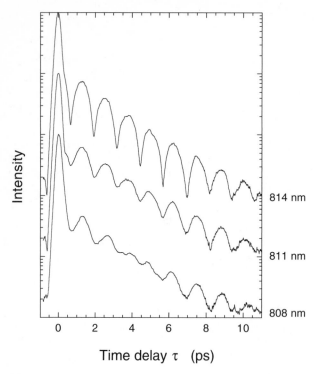

Fig. 2.6. Measured temporal evolution of the microcavity emission at HH exciton resonance when the center energies (wavelengths) of the pump pulses are 1526.26 meV (814 nm), 1531.91 meV (811 nm), and 1537.59 meV (808 nm)

Next we fixed the sample position at the HH exciton resonance and measured the microcavity emission as we tuned the pump laser frequency. When the center frequency of the pump pulses was close to the LH exciton emission frequency, we observed collapse and revival of the oscillations of the microcavity emission (see Fig. 2.6). To explain this phenomenon we calculated the microcavity emission in both the time domain and the frequency domain, as shown in Fig. 2.7. When the center frequency of the pump pulses was tuned toward the LH exciton emission frequency, the LH exciton polariton state (i.e. polariton 3 in Fig. 2.4) was also excited by the pump pulses, besides the two HH exciton polariton states (polaritons 1 and 2). When the magnitude of polariton 3 is almost equal to that of polariton 1, the beating between polaritons 1 and 2 produces an oscillation with a frequency $\omega_h = (E_2 - E_1)/\hbar$, and the beating between polaritons 2 and 3 produces another oscillation with a frequency $\omega_l = (E_3 - E_2)/\hbar$. E_1, E_2, E_3 are the energies of the three polariton states. These two oscillations have almost same amplitude but slightly different oscillation periods. When the two oscillations are 180° out of phase (e.g. at $t = \pi/(\omega_l - \omega_h) \simeq 3.5$ ps), the oscillation of the

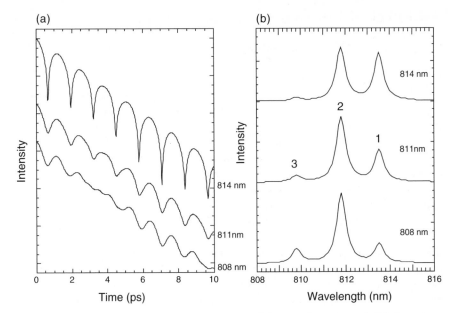

Fig. 2.7. Calculated microcavity emission in (**a**) time domain and (**b**) frequency domain, when the center energies (wavelengths) of the pump pulses are 1526.26 meV (814 nm), 1531.91 meV (811 nm), and 1537.59 meV (808 nm)

microcavity emission collapses. When the two oscillations come into phase again (e.g. at $t = 2\pi/(\omega_l - \omega_h) \simeq 7.0$ ps), the oscillation of the microcavity emission revives.

Note that the collapse and revival of the exciton polariton oscillation observed in a semiconductor microcavity has a different physical origin from the collapse and revival of the vacuum Rabi oscillation in the atom cavity case [14, 70]. The latter is caused by the photon-number-dependent Rabi frequency and the discreteness of the photon number eigenvalues. The origin of this nonlinearity (photon-number-dependent Rabi frequency) lies in the fact that a single two-level atom behaves as a fermion, which allows single excitation. However, in a semiconductor microcavity, the collapse and revival of the exciton polariton oscillation occurs in the completely linear regime, where the exciton density is much lower than the Mott density. The collapse and revival are due to the beating of three exciton polariton modes formed by the strong coupling of both the HH exciton state and the LH exciton state to the cavity photon state.

2.3 Effect of Inhomogeneous Broadening

2.3.1 Exciton Polariton Ladder

In our microcavity samples grown by MBE, the amplitude of the exciton polariton normal-mode splitting agrees well with the theoretical value deduced
from the excitonic oscillator strength in a GaAs QW on the basis of a simple homogeneous system [58,59]. However, in one of our microcavity samples
grown by metalorganic chemical-vapor deposition (MOCVD), the observed
exciton polariton splitting is much smaller [71]. The corresponding temporal
oscillation of the microcavity emission shows a much longer oscillation period; however, the measured oscillation period is consistent with the spectral
splitting observed in this sample. Our study showed that this small splitting
is due to inhomogeneous broadening of QW exciton levels.

As shown in Fig. 2.8, the interfaces of GaAs/AlGaAs QWs consists of
many islands owing to monolayer fluctuations [72]. If the lateral size of the
islands is much smaller than the exciton Bohr radius, the exciton wavefunction is extended over many islands which have different well thickness. This
leads to a large homogeneous broadening due to interface scattering but a relatively small inhomogeneous broadening due to the averaging effect, as shown
schematically in Fig. 2.8a. On the other hand, if the lateral size of the islands
is larger than the exciton Bohr radius, an exciton tends to be localized within
one particular island, and has a narrow homogeneous linewidth due to the
absence of interface scattering but a relatively large inhomogeneous broadening due to the absence of the averaging effect, as shown in Fig. 2.8b. In the
extreme case of a narrow homogeneous linewidth and a large separation of the
discrete exciton energies, multiple discrete exciton states, each corresponding
to a particular well thickness, should appear, as shown in Fig. 2.8b. The energy spacing between those discrete exciton states is equal to the change of
the QW exciton energy due to a one-monolayer change of the well thickness.

Let us consider the coupling of those discrete QW exciton states to the
cavity photon mode in a planar DBR microcavity [73,74]. In the linear coupling regime, where the exciton density is much smaller than the Mott density,
the Hamiltonian of this inhomogeneous system can be written as follows [73]:

$$H = \hbar\omega_c \left(a^\dagger a + \frac{1}{2} \right) + \sum_i \hbar\Omega_i \left(b_i^\dagger b_i + \frac{1}{2} \right) + \sum_i \hbar\Omega_i (a^\dagger b_i + b_i^\dagger a) , \quad (2.30)$$

where b_i is the annihilation operator for the ith exciton, $\hbar\omega_i$ is the energy of
the ith exciton, Ω_i is the coupling constant of the ith exciton to the cavity
photon, and γ_i represents the decay rate of the ith exciton. If $\Omega_i < \gamma_i/2$
and $\gamma_c/2$, the exciton–photon coupling is weak, and thus the cavity is merely
a passive filter for the multiple absorption lines of the QW, as shown by
the dashed lines in Fig. 2.9. As we tune the cavity photon energy, we shift
the transmission window of the bandpass filter, and the absorption spectrum
evolves periodically from three peaks (when the energy of the photon state

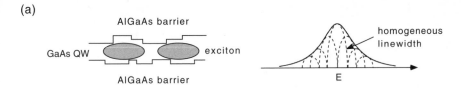

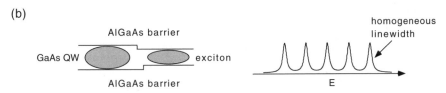

Fig. 2.8. Monolayer fluctuations at the QW interfaces lead to formation of islands. (a) The size of the islands is smaller than the exciton Bohr radius, leading to a single, partially homogeneously broadened exciton state. (b) The size of the islands is larger than the exciton Bohr radius, resulting in multiple, discrete inhomogeneous exciton states

coincides with the energy of one exciton state: point A in Fig. 2.9) to two peaks (when the energy of the photon state is in between the energy of two adjacent exciton states: point B), and back to three peaks (point C). In this case, the energy of each peak does not move, since it corresponds to the energy of the bare exciton state in the QW (see Fig. 2.10a). On the other hand, when $\Omega_i > \gamma_i/2$ and $\gamma_c/2$, the exciton–photon coupling is strong, and thus new eigenmodes (exciton polariton) appear. The exciton polariton dispersion curves, shown as solid lines in Fig. 2.9, feature repeated multiple anticrossing. Since the exciton polariton dispersion curves look like a ladder, this set of curves is called the "exciton polariton ladder". Figure 2.10b shows the calculated absorption spectra in such a strong-coupling regime. Assuming the initial state is the bare photon state and expanding the initial state by means of the exciton polariton states, we can obtain the expansion coefficient c_j for each exciton polariton state. The system energy should be distributed among the exciton polariton modes with a weighting factor $|c_j|^2$. From Fig. 2.10b, we can see that the absorption spectrum evolves periodically from two peaks rather than three peaks (at point A), due to anticrossing, to three peaks in between two anticrossing points (at point B) and back to two peaks (at point C). The unmistakable difference as compared with the weak-coupling regime is that in the strong-coupling regime the energy of each absorption peak shifts with the cavity detuning. For example, when the spectrum evolves from two peaks to three peaks, the new central peak appears exactly in the middle between the previous two peaks.

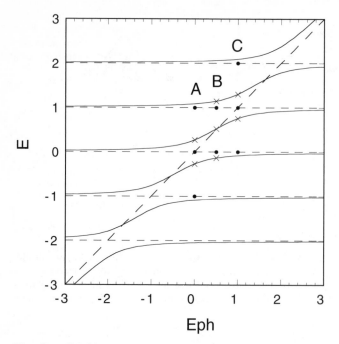

Fig. 2.9. *Solid lines* with *crosses*: exciton polariton dispersion curves in the strong-coupling regime. *Dashed lines* with *filled circles*: bare-exciton and photon dispersion curves in the weak-coupling regime

The microcavity sample used in our experiments described below was grown by MOCVD and consists of a 20 nm GaAs QW in a λ DBR cavity. The top and bottom mirrors consist of 23 and 29.5 pairs, respectively, of $Al_{0.15}Ga_{0.85}As$ and AlAs layers. The $Al_{0.3}Ga_{0.7}As$ spacer layers are tapered along one direction of the sample so that the resonant photon frequency of the cavity varies with sample position. The coupled exciton polariton modes were probed by an absorption measurement at 4.2 K using a mode-locked Ti:sapphire laser. A mode-locked Ti:sapphire laser has a much lower divergence than a white light source. The low divergence of the probe beam was critical in this experiment, because in order to resolve the coupling of the photon mode to an exciton mode, the linewidth of the photon mode must be less than the exciton mode spacing. To reduce the broadening of the cavity photon mode in the tapered cavity used here, the probe beam must be focused to a tiny spot (diameter $\sim$ 15 µm) with a small divergence angle (half angle $\sim 1°$). This could not be achieved with a standard white light source.

As shown in Fig. 2.11, the measured dispersion curves feature multiple anticrossing. This indicates the strong coupling of the multiple QW exciton states to the cavity photon state. Hence, the absorption peaks correspond to absorption by exciton polariton states, rather than uncoupled bare-exciton

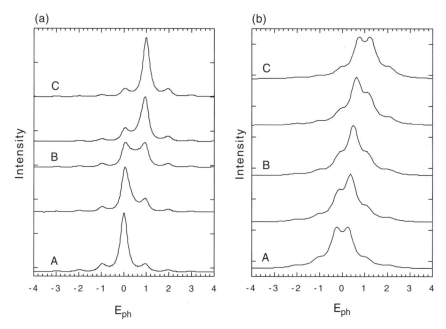

Fig. 2.10. Evolution of the absorption spectra (**a**) in the weak-coupling regime, (**b**) in the strong-coupling regime

states. The measured absorption spectra showed a periodic evolution from two peaks at the anticrossing point (a, c), to three peaks in between two anticrossing points (b, d) [74].

The energy separation between adjacent exciton states was measured to be about 0.3 meV, which agrees well with the change of the QW exciton energy due a one-monolayer ($\sim$ 2.7 Å) change of the QW thickness in a 20 nm GaAs/Al$_{0.3}$Ga$_{0.7}$As QW [34]. Thus, discrete exciton modes result from the monolayer fluctuations of the QW thickness. Since the total excitonic oscillator strength f_T is distributed over many ($N \sim 20$) discrete exciton states, the coupling constant of each QW exciton state to the cavity photon state $\Omega_i \propto \sqrt{f_T/N}$ is significantly reduced, leading to the observed small splitting [74].

2.3.2 Motional Narrowing

Motional narrowing refers to the reduction of the width of a spectral line in a disordered system by some averaging process. A classical particle can be fully localized, giving a transition line shape which simply reflects the probability distribution of the potential. By contrast, the states of a quantum particle may have a finite extent, with the consequence that spatial averaging over the disorder potential causes a reduction in the linewidth [75]. If the lateral

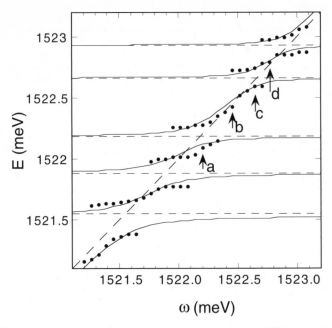

Fig. 2.11. Exciton polariton dispersion curves (*filled circles*) deduced from absorption measurement. The *solid lines* are theoretically fitted exciton polariton dispersion curves. The *dashed lines* are the uncoupled exciton and photon dispersion curves

size of the QW islands is close to the exciton Bohr radius, a single broad exciton line exists in the PL spectrum of a bare QW. However, when the QW is embedded in a high-Q planar microcavity, the spectral linewidth of the polariton is reduced owing to motional narrowing [76, 77]. More specifically, when the QW exciton is dressed by the cavity photon to form the polariton, the effective mass of the polariton is much smaller than that of the exciton, and thus it is much easier for the polariton to move around in the QW plane. Hence the disorder potential is averaged out and the spectral linewidth is reduced. Whittaker et al. have reported the experimental observation of motional narrowing of the lower-polariton linewidth in a high-Q microcavity near resonance [76]. Savona et al. [77] have lifted the transverse-momentum conservation in their calculation of inhomogeneous broadening in a planar microcavity and treated disorder scattering and exciton–photon coupling on an equal footing. Despite the nonconservation of the transverse momentum due to disorder, their numerical simulations have shown that exciton polariton normal-mode splitting is still present in a high-Q planar cavity, and the lower polariton is subject to motional narrowing of the spectral response, while for the upper polariton this effect is less important.

2.4 Effect of Electric and Magnetic Fields

Since electric and magnetic fields can modify excitonic oscillator strengths, they can control the exciton–photon coupling in a microcavity [78–93].

When an electric field is applied perpendicular to the QW plane, the electron in an exciton is pushed against one wall of the well, and the hole against the other wall. Hence the exciton transition energy is lowered. This is the so-called "quantum-confined Stark effect" (QCSE) [94]. In addition, the excitonic oscillator strength is reduced owing to a reduction in the overlap of the wavefunctions of the electron and hole in an exciton. In a QW-embedded microcavity, a perpendicular electric field results in a reduction of the exciton–photon coupling strength, and thus a decrease of the exciton polariton mode splitting [95].

On the other hand, a magnetic field perpendicular to the QW plane shrinks the size of an exciton, and thus enhances the excitonic oscillator strength. In a QW-embedded microcavity, a perpendicular magnetic field increases the exciton–photon coupling strength, and thus increases the exciton polariton mode splitting [96, 97]. Under a high magnetic field ($B > 6.6$ T), Zeeman splitting has been observed on the lower polariton branch [96]. In addition, a very high magnetic field ($B = 11$ T) can enhance the oscillator strength of a high-order light-hole exciton significantly and push its coupling to the cavity photon mode from the weak-coupling regime to the strong-coupling regime [97].

A strong magnetic field has also a significant effect on free carriers in a QW-embedded microcavity [98]. Without a magnetic field, a free electron–hole pair has a continuous energy spectrum. When a continuum of electronic excitations interacts with a single cavity photon mode, Fermi's golden rule applies and the spontaneous-emission process is irreversible. However, a strong magnetic field perpendicular to the QW plane quantizes the in-plane motion of the free carriers into Landau levels and discretizes their density of states. If the cavity mode is between two magneto-optical transitions, it is just like an empty cavity. However, if the cavity photon mode comes into resonance with a magneto-optical transition, the strong-coupling condition is restored, leading to the vacuum Rabi splitting. In the time domain, the spontaneous-emission process becomes reversible. Therefore a magnetic field can continuously tune the interaction between the free carriers and the cavity photon mode from the weak-coupling regime to the strong-coupling regime.

2.5 Transition from Strong to Weak Coupling

At high pump power, the rapid dephasing of excitons and bleaching of excitonic oscillator strength induce a transition from strong coupling to weak coupling [202, 203, 216]. Here we report the direct observation of such a transition in the time domain.

The temporal evolution of the microcavity emission was measured by means of an AC balanced homodyne detection system [99]. Pulses 150 fs long from a mode-locked Ti:sapphire laser were split by a beam splitter into two arms of a modified Mach–Zehnder interferometer. One beam was used as the local-oscillator wave; the other was used to resonantly excite the microcavity sample at an incidence angle of 2.5°. The reflected pulses and the microcavity emission in the reflection direction were combined with the local-oscillator pulses at a second beam splitter. The two outputs from the second beam splitter were detected by two identical photodetectors, whose photocurrents were fed into a differential amplifier. The intensity noise of the local oscillator was reduced by 35 dB owing to the common-mode rejection of the balanced homodyne detector. To eliminate the effect of instability of the interferometer, the optical path length of the signal arm was modulated by Δl at a frequency ν_1 by a mirror mounted on a piezoelectric transducer (PZT) scanner. This optical-path-length modulation generates a sinusoidal signal in the differential-amplifier output at a frequency $\nu_m = \nu_1 \times \Delta l / \lambda$, where λ is the center wavelength of the optical pulses. Since this AC balanced homodyne detection scheme is insensitive to the long-term drift and short-term instability of the Mach–Zehnder interferometer, it gives ultra-high sensitivity. The time delay τ of the local-oscillator pulses could be varied by moving a corner mirror placed on a translational stage. The time evolution of the amplitude of the coherent emission from the microcavity was detected by measuring the sinusoidal output signal at the frequency ν_m at various time delays τ by means of a narrow-bandpass filter and an AC voltmeter .

In Fig. 2.12a we show the temporal evolution of the microcavity emission at different pump powers. The peak at $\tau = 0$ corresponds to the direct reflection of the pump pulses from the sample surface, while the subsequent peaks correspond to the exciton polariton oscillation. In the following analysis, we have used the first peaks as the zero-time markers and exclude them from the evaluation of the decay rate of the microcavity emission. The spot size of the pump beam on the sample was about $25\,\mu$m in diameter. The spectra of the pump pulses were centered at 814 nm (1526.26 meV), with an FWHM of 9 nm (17 meV). The cavity photon frequency was tuned close to the QW heavy-hole exciton emission frequency. The exciton densities were A, 1.1×10^8 cm^{-2}; B, 1.1×10^9 cm^{-2}; C, 5.5×10^9 cm^{-2}; D, 1.1×10^{10} cm^{-2}; E, 2.0×10^{10} cm^{-2}; F, 2.7×10^{10} cm^{-2}; G, 4.4×10^{10} cm^{-2}; H, 6.6×10^{10} cm^{-2}; I, 1.1×10^{11} cm^{-2}. In Fig. 2.12b we show a simultaneous measurement of the reflection spectra of the probe beam. At low pump power, the strong coupling of the QW HH exciton with the cavity photon results in two HH exciton polariton peaks in the reflection spectrum. The beating between these two exciton polariton states leads to a temporal oscillation of the microcavity emission. The oscillation period is about 1.2 ps, which is in good agreement with the spectral splitting of about 2.1 nm (4.1 meV) between the two HH exciton polariton states. As we increase the excitation intensity, in the fre-

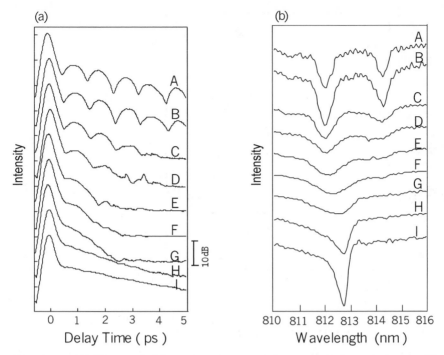

Fig. 2.12. (a) Measured temporal evolution of the microcavity emission at different pump powers. (b) Simultaneous measurement of the reflection spectra from the microcavity at different pump powers. The estimated exciton densities in the microcavity are A, 1.1×10^{8} cm^{-2}; B, 1.1×10^{9} cm^{-2}; C, 5.5×10^{9} cm^{-2}; D, 1.1×10^{10} cm^{-2}; E, 2.0×10^{10} cm^{-2}; F, 2.7×10^{10} cm^{-2}; G, 4.4×10^{10} cm^{-2}; H, 6.6×10^{10} cm^{-2}; I, 1.1×10^{11} cm^{-2}

quency domain the exciton polariton peaks are broadened, and their mode splitting is slightly reduced. In the time domain, the microcavity emission peak intensity decays faster, but the exciton polariton oscillation becomes slower. Eventually the two exciton polariton peaks in the reflection spectra are replaced by a single cavity photon peak, indicating the transition from strong exciton–photon coupling to weak coupling. The single cavity photon resonance narrows as the pump power increases. Meanwhile, the temporal oscillation of the microcavity emission is replaced by an exponential decay. The decay rate decreases as the pump power increases.

In Fig. 2.13 we show the measured decay rate and peak wavelength in the reflection spectrum of the microcavity emission as a function of the pump power. Since the pump pulse duration is much shorter than the decay time of the microcavity emission in the strong-coupling regime, we extracted the exciton polariton decay rate from the slope of the oscillation peaks in Fig. 2.12a, i.e. by fitting the height of the oscillation peaks with an exponential function.

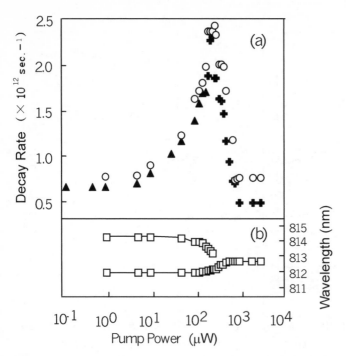

Fig. 2.13. Measured decay rate and peak wavelength in the reflection spectrum of the microcavity emission as a function of the pump power. (**a**) Measured decay rate. (**b**) Measured peak wavelength in the reflection spectrum

In the weak-coupling regime, we obtained the emission decay rate by fitting the intensity with an exponential function.

In Fig. 2.13a, the triangles represent the decay rate in the strong-coupling regime at low pump power, while the crosses represent the decay rate in the weak-coupling regime at high pump power. The circles represent the decay rate γ_T evaluated from the spectral linewidth $\Delta\omega$ of the reflection spectrum by using the relation $\gamma_T = \Delta\omega/2$. It is clear that the decay rates evaluated from the temporal evolution are in close agreement with those obtained from the corresponding spectral linewidth. As the pump power increases, the decay rate of the exciton polariton oscillation increases. At the transition point from strong coupling to weak coupling, the decay rate is maximum. After passing the transition point, as the pump power increases further, the exponential decay rate of the microcavity emission starts decreasing. In Fig. 2.13b we show the peak wavelength in the reflection spectrum as a function of the pump power.

In order to understand the experimental results, we consider the quantum mechanical model of two coupled oscillators in contact with reservoirs which introduce damping into the system, as developed in Sect. 2.1.2. Using the solutions of the Heisenberg equations and the initial condition of a bare

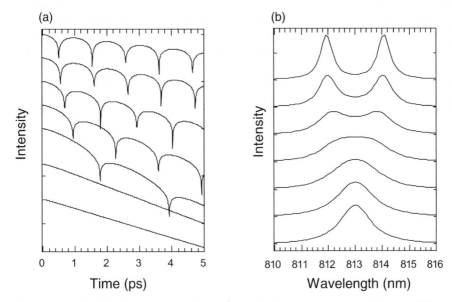

(a)

Intensity

0   1   2   3   4   5

Time (ps)

(b)

Intensity

810 811 812 813 814 815 816

Wavelength (nm)

Fig. 2.14. (a) Calculated temporal evolution of the microcavity emission at different values of Γ_h. (b) Calculated absorption spectra at different values of Γ_h. From *top* to *bottom*, the values of Γ_h are 1, 2, 4, 6, 8, 10, and 12 meV

photon, it is straightforward to calculate the microcavity emission, which is shown in Fig. 2.14a. When $\gamma/2 < \Omega$, the microcavity emission shows a temporal oscillation. This indicates that the microcavity system oscillates back and forth between the QW exciton state and the cavity photon state. As γ increases, the oscillation decays faster, because the decay rate of the exciton polariton emission is $(1/2)(\gamma + \gamma_c)$. The oscillation period, which is inversely proportional to the exciton polariton mode splitting Ω, increases slightly owing to the slight decrease in Ω at larger γ. When $\gamma/2$ approaches Ω, the temporal oscillation is replaced by an exponential decay. This change in the time domain is accompanied by the merging of the two exciton polariton peaks into a single peak in the frequency domain. As γ increases further, the decay rate of the microcavity emission starts decreasing, and eventually it approaches γ_c in the weak-coupling limit. The calculation results agree well with the experimental data. We used $\Omega = 4.1$ meV, $\gamma_c = 1$ meV, on the basis of experimental parameters of our samples.

Concerning the frequency domain, we show in Fig. 2.14b the calculated absorption spectra as functions of γ. As γ increases, the exciton polariton peaks are broadened, because the linewidth of an exciton polariton peak is proportional to the sum of the QW exciton linewidth (γ) and the cavity photon linewidth, i.e. $(\gamma_c + \gamma)/2$. The splitting Ω also decreases owing to the increase of γ, according to (2.20). When $\gamma/2$ approaches Ω, the two exciton polariton peaks merge into a single broad peak, indicating the transition from strong coupling to weak-coupling. As γ increases further, the linewidth of this single peak decreases, eventually approaching the bare-photon resonance linewidth γ_c. This is because in the weak-coupling regime, when $\Gamma \gg \Gamma_c$, the linewidth of the absorption peak is determined mostly by Γ_c. Therefore, our simulation results are consistent with our experimental reflectivity data.

3. Biexcitonic Effects in Microcavities

3.1 Excitons Versus Two-Level Atoms

All the experiments described in Chap. 2 were performed in the low-excitation regime. What has been observed in a semiconductor microcavity is similar to what has been observed in an atomic cavity. So the question is: is there any difference between semiconductor cavity QED and atomic cavity QED? To answer this question, let us compare a dressed atom with a dressed exciton.

Figure 3.1 shows the dressed-state level diagrams of the $N = 0, 1, 2$ excitation manifolds. The dressed states are the eigenstates of the total Hamiltonian, which includes the light–matter interaction term, while the bare states are the eigenstates of a Hamiltonian which does not include the light–matter interaction term. Let us consider a single two-level atom inside a cavity. When there is only one excitation, there are two bare states: one is a state in which the atom is in the ground state and one photon exists in the cavity mode, $|g, 1_{ph}\rangle$; the other is a state in which the atom absorbs the photon and goes to the excited state and there is no photon in the cavity, $|e, 0_{ph}\rangle$. In the resonant case, i.e. where the cavity photon energy is equal to the energy difference between the excited state and ground state of the atom, these two bare states, $|g, 1_{ph}\rangle$ and $|e, 0_{ph}\rangle$, have the same energy. However, the strong coupling of the atom to the cavity photon mode mix the two bare states and forms two dressed states, $|1_+\rangle$ and $|1_-\rangle$, with an energy splitting $2g$, where g is the atom–photon coupling constant. This gives the vacuum Rabi splitting. However, in an imperfect cavity, dressed states are not stable, because photons can leak out of the cavity. The decay of the cavity photon results in transitions from the two dressed states in the $N = 1$ manifold to the ground state $|g, 0_{ph}\rangle$. These two transitions result in two peaks in the spectrum of the light leaking out of the cavity.

This situation is similar in a semiconductor microcavity. When there is one excitation, there are two bare states: one is a state in which there is one photon in the cavity but there is no exciton, $|0_{ex}, 1_{ph}\rangle$; the other is a state in which the photon is absorbed to create an exciton and thus there is no photon in the cavity, $|1_{ex}, 0_{ph}\rangle$. In the resonant case, i.e. where the cavity photon energy is equal to the exciton energy, these two bare states, $|0_{ex}, 1_{ph}\rangle$ and $|1_{ex}, 0_{ph}\rangle$, are degenerate in energy. However, the strong exciton–photon coupling lifts the energy degeneracy of these two bare states and results in

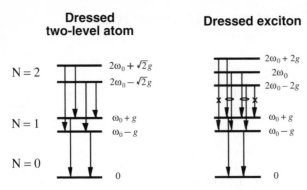

Fig. 3.1. Dressed-state level diagrams of the $N = 0, 1, 2$ excitation manifolds. The forbidden transitions are marked with *crosses*, and the two transitions which have the same frequency are *circled* together

two new eigenstates $|1_+\rangle$ and $|1_-\rangle$, i.e. exciton polaritons, with an energy splitting $2g$, where g is the exciton–photon coupling constant. Exciton polaritons are not stable states, because photons can leak out of the cavity and excitons can be scattered into other momentum states and also recombine nonradiatively. The decay of the cavity photon and exciton results in the decay of the exciton polariton, and leads to two peaks in the microcavity emission spectrum. Therefore the exciton–photon coupled system in a semiconductor microcavity behaves similarly to the atom–photon coupled system in an atomic cavity in the low-excitation regime.

However, as we increase the excitation level, a difference between dressed atoms and dressed excitons appears. In the case of a single two-level atom inside a cavity, when there are two excitations, there are still only two bare states: one is a state in which the atom is in the ground state and there are two photons in the cavity, $|g, 2_{ph}\rangle$; the other is a state in which the atom absorbs one photon and goes to the excited state and there is one photon in the cavity, $|e, 1_{ph}\rangle$. The atom–photon coupling lifts the energy degeneracy of these two bare states and results in two new eigenstates $|2_+\rangle$ and $|2_-\rangle$. However, the energy separation between $|2_+\rangle$ and $|2_-\rangle$ is different from the energy separation between $|1_+\rangle$ and $|1_-\rangle$. The decay of the cavity photon results in transitions from the dressed states in the $N = 2$ manifold to the dressed states in the $N = 1$ manifold, and subsequently from the dressed states in the $N = 1$ manifold to the ground state. But the four transitions from $|2_+\rangle$ and $|2_-\rangle$ to $|1_+\rangle$ and $|1_-\rangle$ have different frequencies from the two transitions from $|1_+\rangle$ and $|1_-\rangle$ to $|g, 0_{ph}\rangle$. Therefore they give additional peaks in the emission spectrum.

On the other hand, in a semiconductor microcavity, when there are two excitations, there are three bare states: one is a state in which there are two photons in the cavity and no excitons, $|0_{ex}, 2_{ph}\rangle$; another is a state in which one photon is absorbed to create an exciton and there is one photon in the cavity, $|1_{ex}, 1_{ph}\rangle$; the third one is a state in which two photons are absorbed to create two excitons and there is no photon in the cavity, $|2_{ex}, 0_{ph}\rangle$. Thus the exciton–photon coupling results in three dressed states $|2_+\rangle$, $|2_0\rangle$, and $|2_-\rangle$. But these three dressed states are equally spaced in energy. The energy separation between them is equal to the energy separation between the two dressed states in the $N = 1$ manifold. The decays of the cavity photon and of the exciton result in transitions from the three dressed states in the $N = 2$ manifold to the dressed states in the $N = 1$ manifold, and subsequently from the dressed states in the $N = 1$ manifold to the ground state. Although there are six possible transitions from $|2_+\rangle$, $|2_0\rangle$, and $|2_-\rangle$ to $|1_+\rangle$ and $|1_-\rangle$, the two outlying transitions (marked with crosses in Fig. 3.1) are forbidden, and the remaining four transitions are pairwise degenerate, with the transition frequencies coinciding with those of the two transitions from $|1_+\rangle$ and $|1_-\rangle$ to the ground state. Therefore the emission spectrum still consists of only two peaks. This is easy to understand, because excitons behave like bosons. Hence, the coupling between the exciton and photon is just like the coupling between two linear harmonic oscillators, which always features two constant eigenfrequencies regardless of the excitation level. On the other hand, two-level atoms behave like fermions. Hence, at higher excitation intensity, saturation brings nonlinearity into the atom–photon system and results in additional peaks in the emission spectrum.

From the mathematical point of view, the Hamiltonian of the coupled atom–photon system is

$$H = \hbar\omega_c\left(a^\dagger a + \frac{1}{2}\right) + \hbar\omega_a\left(b_a^\dagger b_a + \frac{1}{2}\right) + \hbar g(a^\dagger b_a + ab_a^\dagger)\,, \tag{3.1}$$

where a and b_a are the operators for the cavity photon and the two-level atom, respectively, $\hbar\omega_c$ is the energy of the cavity photon, $\hbar\omega_a$ is the energy difference between the excited state and ground state of the atom, and g is the atom–photon coupling constant. This Hamiltonian is similar to that of the exciton–photon coupled system (2.11). However, the commutation relation for the atom operator b_a is different from the commutation relation for the exciton operator b, i.e.

$$\{b_a, b_a^\dagger\} = b_a b_a^\dagger + b_a^\dagger b_a = 1\,, \tag{3.2}$$
$$[b, b^\dagger] = bb^\dagger - b^\dagger b = 1\,. \tag{3.3}$$

Therefore, the Heisenberg equation of motion for b_a, which is derived from from (3.1), is different from that of b in (2.16):

$$\frac{da}{dt} = -i\omega_c a - igb_a - \frac{\Gamma_c}{2}a\,, \tag{3.4}$$
$$\frac{db_a}{dt} = -i\omega_a b_a - iga - \frac{\Gamma_a}{2}b_a + 2b_a^\dagger b_a(i\omega_a b_a + iga)\,. \tag{3.5}$$

The last term $2b_a^\dagger b_a(i\omega_a b_a + iga)$ in (3.5) is a nonlinear term. It is proportional to the mean occupancy of the atom, $\bar{n} = \langle b_a^\dagger b_a\rangle$. Therefore the solutions of

(3.4) and (3.5), $a(t)$ and $b_a(t)$, depend on the excitation level $\bar{n}$. This nonlinear term is absent in (2.16) for the exciton–photon coupled system, owing to the bosonic commutation relation of the exciton operator b (3.3). Since (2.15) and (2.16) are linear, the solutions $a(t)$ and $b(t)$ are independent of the excitation level. This is consistent with the conclusion drawn from our previous analysis.

3.2 Exciton–Exciton Interaction

So, in the nonlinear regime, we can see the difference between an exciton–photon coupled system and an atom–photon coupled system. But things are not so simple. Up to now we have assumed the excitons did not interact with each other. This is true only when the exciton density is low and the spatial separation between excitons is large.

As mentioned in Sect. 2.4, monolayer fluctuations of the quantum well thickness result in many islands. If the lateral size of those islands is equal to or larger than the exciton Bohr radius, an exciton tends to localize within one island. As the excitation intensity increases, the probability of two excitons occupying the same island increases. If the lateral size of the island is comparable to the exciton Bohr radius, each island can only hold one exciton, owing to the repulsive Coulomb interaction between excitons at such short distances. Hence, such an exciton system behaves more like a fermionic system. However, in a larger island, the distance between the excitons can be substantial, and the Coulomb repulsion between excitons is smaller. In sufficiently large islands, the attractive interaction between two excitons with opposite angular momentum is dominant, leading to the formation of biexcitons. Such an attractive interaction favors an occupation of more than one exciton in one island. In this case the exciton system is more like a bosonic system. Finally, if the island size is much larger than the exciton Bohr radius, the large distance between excitons makes the exciton–exciton interaction negligibly small, and thus excitons can be approximated as noninteracting bosons. Hence, from the quantum statistical point of view, a QW exciton system is a very interesting system to study because it can demonstrate a continuous transition from a system of fermions to a system of bosons. Next, we shall show that the exciton–exciton interaction also results in the onset of additional spectral sidebands in the exciton–photon coupled system. This indicates a fundamental connection between a system of fermions (two-level atoms) and a system of interacting bosons (excitons) at high densities.

The system Hamiltonian including the exciton–exciton interaction term can be written as follows:

$$\hat{H} = \sum_{\sigma} \hbar\omega_p \hat{a}_{\sigma}^{\dagger}\hat{a}_{\sigma} + \sum_{\sigma} \hbar\omega_{ex}\hat{b}_{\sigma}^{\dagger}\hat{b}_{\sigma} + \sum_{\sigma} \hbar g(\hat{a}_{\sigma}^{\dagger}\hat{b}_{\sigma} + \hat{a}_{\sigma}\hat{b}_{\sigma}^{\dagger})$$

$$+ \sum_{\sigma,\sigma'} \hbar B_{\sigma\sigma'}(r)\hat{b}_{\sigma}^{\dagger}\hat{b}_{\sigma'}^{\dagger}\hat{b}_{\sigma'}\hat{b}_{\sigma} , \tag{3.6}$$

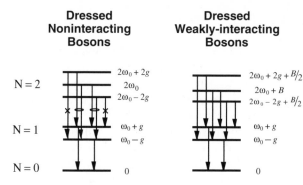

Fig. 3.2. Dressed-state level diagram of $N = 0, 1, 2$ excitation manifolds for excitons considered as weakly interacting bosons and as noninteracting bosons. The level energies are given to the lowest order of B. The forbidden transitions are marked with *crosses*, and the two transitions which have the same frequency are *circled* together

where $\hat{a}_\sigma$ and $\hat{b}_\sigma$ are the photon and exciton operators, respectively, which obey the bosonic commutation relation, σ represents the polarization or angular momentum J_z of the photon or exciton, respectively, $\hbar\omega_p$ and $\hbar\omega_{\mathrm{ex}}$ are the energies of the bare photon and exciton states, respectively, and g is the coupling constant between the two. $B_{\sigma\sigma'}$ is the exciton–exciton interaction coefficient, whose value depends on the distance between the excitons r. The interaction between two excitons with opposite J_z is repulsive at small r and attractive at large r, while the interaction between two excitons with the same J_z is always repulsive [183]. Hence, depending on the average distance $\bar{r}$ between excitons (which is determined by the exciton density and QW island size), the dominant exciton–exciton interaction can be either attractive (at large $\bar{r}$) or repulsive (at small $\bar{r}$).

We can then solve for the eigenstates and eigenvectors of the total Hamiltonian (3.6). The decay of the bare photon and bare exciton states results in transitions between the excitation manifolds $N + 1$ and N with transition amplitudes $\langle N|\hat{a}|N+1\rangle$ and $\langle N|\hat{b}|N+1\rangle$, respectively [101]. At densities where excitons behave like weakly interacting bosons, the energies of the three dressed exciton states in the $N = 2$ manifold are shifted from those of noninteracting bosons (see Fig. 3.2). The exciton–exciton interaction switches on the two forbidden $2 \to 1$ transitions from the dressed states in the $N = 2$ manifold to the dressed states in the $N = 1$ manifold. It also lifts the frequency degeneracies in the other four $2 \to 1$ transitions, and red-shifts or blue-shifts the $2 \to 1$ transition frequencies depending on whether the exciton–exciton interaction is mainly attractive or repulsive. Therefore the exciton–exciton interaction results in additional peaks in the emission spectrum.

3.3 Biexcitonic Effects

As mentioned in Sect. 2.4, in one of our microcavity samples grown by MOCVD, we have observed multiple, discrete exciton lines in the absorption and PL spectra. Those discrete QW exciton states result from the monolayer fluctuations of the well width. An image of the QW structure taken with a high-resolution transmission electron microscope (HRTEM) confirmed fluctuations of the well width of many monolayers. It also showed that the lateral size of a typical QW island is around 60 nm. Therefore, when two excitons occupy one island, the interaction between them is attractive if they have opposite J_z, while the interaction is negligibly small if they have the same J_z.

Experimentally, we fixed the excitation spot where the cavity photon state was in resonance with one localized QW exciton state. We resonantly created the localized QW excitons using a 200 fs mode-locked Ti:sapphire laser at an incidence angle of 4° from the normal. The excitation spot was approximately 20 μm in diameter. The sample was cooled to 4.2 K. Since the cavity linewidth (~ 0.2 meV) was narrower than the energy separation between localized QW exciton states, all other localized exciton states were greatly detuned, and could not be excited by the pump light, which was filtered by the cavity. The emission from the cavity in the direction normal to the QW plane was measured by a spectrometer and a power meter. From the emission intensity, we can infer the initial density of excitons created in the QW. Figure 3.3 shows the evolution of the emission spectra as we increased the intensity of the linearly polarized pump beam. When the excitation is weak, the system is excited mainly to the $N = 1$ manifold. In other words, at most only one exciton is created in each of those islands whose QW exciton energy coincides with the resonant photon energy of the cavity . The transition from the $N = 1$ to the $N = 0$ manifold results in a double-peaked emission spectrum. As we increase the excitation intensity, the emission spectrum becomes asymmetric, i.e. the low-frequency peak (peak 2 in Fig. 3.3) becomes higher than the high-frequency peak (peak 1). Meanwhile, two additional peaks (peaks 3 and 4) emerge on the low-frequency side. This indicates that the exciton–exciton interaction is mainly attractive.

To understand the experimental data quantitatively, we have performed a simple numerical simulation of the emission spectra. Since we monitored the photons leaking out of the cavity mode in the normal direction, the measured emission spectra should correspond to the transitions caused by the decay of the bare cavity photon state ($\langle N|\hat{a}|N+1\rangle$). The initial state is a bare exciton state since, experimentally, the pump light is incident on the sample at a small angle, creating excitons with small $k_\parallel$. Those excitons subsequently relax to the $k_\parallel \simeq 0$ state and then strongly couple to the observed cavity mode. Because the lateral size of the QW islands is around 60 nm, in the intermediate excitation regime where the number of excitons created in each island is 1 or 2, we can assume $B_{\sigma=\sigma'} \simeq 0$ and $B_{\sigma\neq\sigma'} \simeq -E_B/2$, where E_B

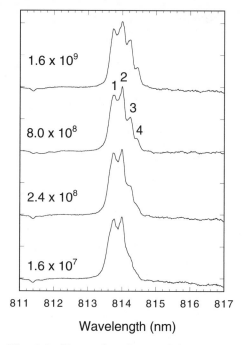

Fig. 3.3. Observed evolution of the emission spectra as the intensity of the linearly polarized pump light was increased. The initial exciton densities created in the QW by the short laser pulses are given in units of cm^{-2}

is the biexciton binding energy in the QW. Figure 3.4 shows the calculated emission spectra of the $2 \to 1$ transitions (solid lines) and the subsequent $1 \to 0$ transitions (dotted lines). From it, we can clearly see that the biexcitonic effect on the $2 \to 1$ transitions is a red shift of the transition frequencies, the emergence of sidebands due to the switch-on of forbidden transitions, and asymmetry of the emission spectrum. Although the subsequent $1 \to 0$ transitions do not shift in frequency, the biexcitonic effect also induces an asymmetry in the emission spectrum because the nonlinear $2 \to 1$ transitions populate the lower energy level in the $N = 1$ manifold more.

Comparing with our simulation, we can explain our experimental data in detail: as the excitation intensity increases, in some islands two excitons are created, that is, the system is excited to a mixture of $N = 1$ and $N = 2$ manifolds. For an island occupied by two excitons, the $2 \to 1$ transitions and the subsequent $1 \to 0$ transitions are affected by the biexcitonic effect. The red shift of the $2 \to 1$ transition frequencies results in peak 3 in Fig. 3.3, and the switch-on of the forbidden $2 \to 1$ transitions results in peak 4. The subsequent $1 \to 0$ transitions lead to an asymmetry in the height of peaks 1 and 2. As the excitation intensity increases further, more islands are occupied by two excitons, and hence the biexcitonic effect becomes more pronounced.

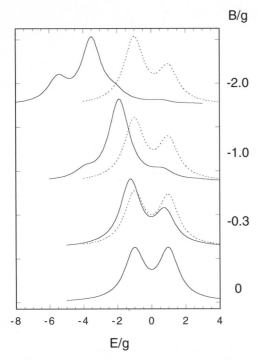

Fig. 3.4. Calculated emission spectra corresponding to transitions from the $N = 2$ to the $N = 1$ manifold (*solid lines*) and the subsequent transitions from the $N = 1$ to the $N = 0$ manifold (*dotted lines*) at different values of B/g

From the measured dispersion curve, we have deduced the coupling constant g between the cavity photon state and the localized QW exciton state to be about 0.24 meV. By measuring how much the $2 \rightarrow 1$ transition frequencies are red-shifted in the emission spectra (Fig. 3.3), and comparing with the simulation (Fig. 3.4), we have deduced that $B \simeq -0.43$ meV. This corresponds to a QW biexciton binding energy $E_B \simeq 0.86$ meV, which is in good agreement with the value inferred from a PL measurement [102].

For a confirmation of the biexcitonic effect, we also measured the emission spectra when the pump beam was circularly polarized. Since all the excitons created resonantly by circularly polarized pump light have the same J_z, the biexcitonic effect is switched off. Experimentally, as we rotated a quarter-wave plate to change the pump beam polarization from linear to circular while keeping the same pump intensity, the emission spectrum changed drastically from four peaks to two peaks (see Fig. 3.5). This confirms that, although in some islands two excitons of the same J_z have been created, the repulsive interaction between them is negligibly small owing to the large island size. Therefore the $2 \rightarrow 1$ transitions give the same double peaks in the emission spectrum as the $1 \rightarrow 0$ transitions. In fact, we did not observe any change in

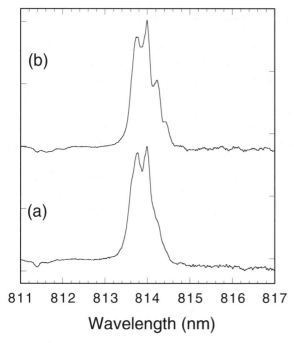

Fig. 3.5. Observed emission spectra when the pump light is (**a**) circularly polarized, (**b**) linearly polarized. The QW exciton density was kept constant in the two cases, i.e. 8×10^8 cm^{-2}. The emission spectra were taken at the same sample position as in Fig. 3.3

the emission spectra under circularly polarized pump light as we varied the pump intensity over the same range as we used with the linearly polarized pump light to obtain the results shown in Fig. 3.3. This confirms that the additional features which appear as we increase the intensity of the linearly polarized pump light indeed originate from the biexcitonic effect.

Finally, we would like to comment on why the biexcitonic effect had not been observed before in a QW-embedded microcavity [103–105], even though biexciton transitions had been observed in bare QWs by many groups [106, 107]. In the case of a microcavity, as shown in Fig. 3.4, the ratio $|B/g|$ describes the relative influence of the biexcitonic effect on the emission spectra. In ordinary MBE-grown microcavity samples, the QW exciton line is homogeneously broadened, and the exciton–photon coupling constant g is of order 1.5 meV. Since B (~ 0.5 meV) is much smaller than g, the biexcitonic effect is weak and difficult to observe. However, in our MOCVD-grown microcavity sample, since the excitonic oscillator strength is distributed over many discrete exciton states, the coupling constant of the cavity photon state with each QW exciton state is reduced significantly (~ 0.2 meV). The drastic increase of $|B/g|$ makes the biexcitonic effect more pronounced and easier to observe in this sample.

In summary, we have shown that the exciton–exciton interaction results in additional spectral sidebands in the exciton–photon coupled system. This indicates a fundamental connection between a system of fermions (two-level atoms) and a system of weakly interacting bosons (excitons) at high densities. Our experimental results can be well explained by our theoretical model. Note that in our model we have neglected other nonlinearities because their effects are very small compared with the biexcitonic effect in the medium-exciton-density regime which we studied in our experiments. As the exciton density increases further, many other nonlinear effects become important. For example, the phase-space-filling effect will induce a reduction of the excitonic oscillator strength. Recently, Gonogami et al. have studied the combination of the biexcitonic effect and the phase-space-filling effect by four-wave-mixing spectroscopy [108].

4. Resonant Tunneling into Exciton and Polariton States

All of the studies on semiconductor cavity QED described so far have been carried out by means of optical pumping. However, electrical pumping is crucial for practical applications. The usual electrical pumping scheme for QW excitons is indirect, namely electrons and holes are injected into a QW and then relax to form excitons. This process is not only slow, but also inefficient. A direct and efficient electrical pumping scheme for QW excitons is desired.

In addition, excitonic effects in the resonant tunneling of photo-excited carriers have attracted much interest because of their fundamental quantum mechanical aspects and potential application to tunneling devices [109, 110]. The tunneling of free electrons (or holes) through a thin barrier between two adjacent QWs has been shown to be a transfer from a direct (intrawell) exciton to an indirect (interwell) exciton [111, 112]. Recently, Lawrence et al. demonstrated that excitons can tunnel as a single entity between CdTe/CdMnTe and CdTe/CdZnTe asymmetric double QWs [113]. However, in those cases the excitons already exist before tunneling. Can we *create* excitons in a QW directly through resonant tunneling?

We have proposed and demonstrated a new resonant tunneling process to directly create excitons in a QW [114].

4.1 Resonant Tunneling into QW Exciton States – Theory

4.1.1 Hole-Assisted Resonant Tunneling of Electrons into QW Exciton States

As shown in Fig. 4.1, the device structure used in our experiments consists of a p-doped $Al_{0.3}Ga_{0.7}As$ layer, an undoped GaAs QW, an undoped $Al_{0.3}Ga_{0.7}As$ barrier, and an n-doped GaAs layer. Without bias, the free holes in the p-doped $Al_{0.3}Ga_{0.7}As$ layer can thermally diffuse into the QW, while the free electrons in the n-doped GaAs layer are blocked by the intrinsic $Al_{0.3}Ga_{0.7}As$ barrier. As we apply a positive bias, the free electrons in the conduction band of the n-doped GaAs layer approach an exciton energy level inside the QW, which is lower than the first electron–hole subband energy level by the exciton

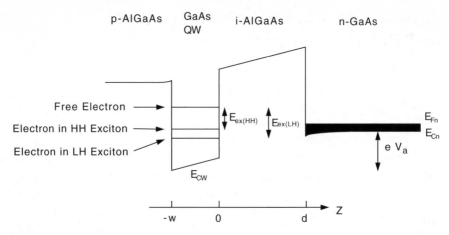

Fig. 4.1. Band structure of our tunneling device under forward bias

binding energy E_{ex}. Then the free electron can tunnel into the QW and combine with a hole to form an exciton directly. Such tunneling is resonantly enhanced when the energy of the initial state (a free electron in the n-doped GaAs layer and a subband hole in the QW) is equal to the energy of the final state (an exciton in the QW).

Let us derive this two-particle resonant-tunneling condition. The initial state is a free electron in the n-doped GaAs layer with a transverse wavevector $\boldsymbol{k}_{e\parallel}$ and a longitudinal wavevector k_{ez}, plus a subband hole inside the QW with a continuous transverse wavevector $\boldsymbol{k}_{h\parallel}$ and a quantized longitudinal wavevector k_{hz} or energy E_{hz}. The final state is a 2D exciton, which is characterized by its center-of-mass transverse wavevector $\boldsymbol{k}_{ex\parallel}$, the internal exciton binding energy E_{ex}, and the longitudinal energy given by the subband energy levels of the electron, E_{ez}, and of the hole, E_{hz}. Since the potential barrier is translationally invariant along the transverse direction, the transverse momentum is conserved during the tunneling, i.e.

$$\boldsymbol{k}_{e\parallel} + \boldsymbol{k}_{h\parallel} = \boldsymbol{k}_{ex\parallel} \,. \tag{4.1}$$

However, the transverse kinetic energy is not necessarily conserved, i.e.

$$\frac{\hbar^2 k_{e\parallel}^2}{2m_e} + \frac{\hbar^2 k_{h\parallel}^2}{2m_{h\parallel}} - \frac{\hbar^2 k_{ex\parallel}^2}{2m_{ex}} = \frac{\hbar^2 k_r^2}{2\mu} \neq 0 \,, \tag{4.2}$$

where $m_{h\parallel}$ is the transverse effective mass of the subband hole, $m_{ex} = m_e + m_{h\parallel}$ is the total mass of the 2D exciton, $\mu = \left(m_e^{-1} + m_{h\parallel}^{-1}\right)^{-1}$ is its internal reduced mass, and $\boldsymbol{k}_r$ is determined by the relative electron–hole transverse velocity:

$$\frac{\boldsymbol{k}_r}{\mu} = \frac{\boldsymbol{k}_{e\parallel}}{m_e} - \frac{\boldsymbol{k}_{h\parallel}}{m_{h\parallel}} \,. \tag{4.3}$$

However, the total energy must be conserved, i.e.

$$\left(\frac{\hbar^2 {k_e}_\|^2}{2m_e + \frac{\hbar^2 {k_{ez}}^2}{2m_e}}\right) + \left(\frac{\hbar^2 {k_h}_\|^2}{2m_{h\|}} + E_{hz}\right) + eV_a$$

$$= \frac{\hbar^2 {k_{ex}}_\|^2}{2m_{ex}} + E_{ez} - E_{ex} + E_{hz} \, , \tag{4.4}$$

where $V_a = V - V_b$, V is the applied forward bias, and V_b is the built-in voltage of the pn junction.

Since the energy difference between the hole subbands inside the QW is much larger than the kinetic energy of a hole in an exciton, and similarly for the electron subbands, we neglect the probability of a hole transition from one subband to another. From the conservation of total energy and transverse momentum, we obtain the resonant-tunneling condition:

$$\frac{\hbar^2 {k_{ez}}^2}{2m_e} = E_{ez} - E_{ex} - eV_a - \frac{\hbar^2 {k_r}^2}{2\mu} \, . \tag{4.5}$$

This condition is different from the conventional one-particle resonant-tunneling condition. For example, when an electron tunnels into an electron subband level inside a QW, the resonant-tunneling condition derived from the conservation of transverse momentum and total energy is

$$\frac{\hbar^2 {k_{ez}}^2}{2m_e} = E_{ez} - eV_a \, . \tag{4.6}$$

This condition implies that all electrons with a certain k_{ez} can resonantly tunnel into the QW, regardless of their initial transverse momentum $\hbar k_{e\|}$. Since resonant tunneling into an exciton level is a two-particle problem, the required k_{ez} for an electron to tunnel in depends on which hole state the electron will combine with to form an exciton. In other words, $k_{e\|}$ and $k_{h\|}$ determine the required k_{ez}. Therefore different initial electron states could tunnel into the same final exciton state, but with different initial hole states. This results in a broader resonant-tunneling peak in a current–voltage (I–V) curve in comparison with the peak from electron tunneling into electronic subband.

As an example, let us consider the upper limit V_{max} and lower limit V_{min} of the effective bias V_a at which the resonant tunneling process can occur. At $T = 0$ K, for resonant tunneling of an electron into an electron level in the QW,

$$V_{max} = E_{ez}/e; \; V_{min} = (E_{ez} - E_{fe})/e,$$

where E_{fe} is the Fermi energy of the electrons with respect to the conduction band edge in the n-doped GaAs layer. On the other hand, for resonant tunneling of an electron into an exciton level in the QW,

$$V_{\text{max}} = (E_{ez} - E_{ex})/e ,$$

$$V_{\text{min}} \tag{4.7}$$

$$= \begin{cases} (E_{ez} - E_{ex} - E_{fe} - |E_{fh}|)/e , & \text{if } k_{fe} > k_{fh} , \\ [E_{ez} - E_{ex} - E_{fe} + E_{fh} + \hbar^2(k_{fh} - k_{fe})^2/2m_{ex}]/e , & \text{if } k_{fe} < k_{fh} , \end{cases}$$

where E_{fh} is the Fermi energy of the holes with respect to the valence band edge inside the QW, and k_{fe} and k_{fh} are the Fermi $\boldsymbol{k}$ vectors of electron and hole, respectively, i.e. $k_{fe} = \sqrt{2m_e E_{fe}}$, $k_{fh} = \sqrt{2m_{h\parallel}|E_{fh}|}$. Below we consider only the case where $k_{fe} > k_{fh}$.

From the above equations, we can see that the range of variation of the bias voltage for the tunneling of an electron into the exciton level is $(E_{fe} + |E_{fh}|)/e$, while for the tunneling of an electron into the electron level the range of variation of the bias is only E_{fe}/e.

4.1.2 Conventional Resonant Tunneling of Electrons

In this section, we briefly review the calculation of the one-particle resonant-tunneling current density. Consider an electron tunneling resonantly from the n-doped GaAs layer through the barrier into the electron subband level in the QW. This tunneling process can be treated as a transition from an initial state $|i\rangle = |\boldsymbol{k}_{e\parallel}, k_{ez}\rangle$ to a final state $|f\rangle = |\boldsymbol{k}'_{e\parallel}, E_{ez}\rangle$ [115]. According to Fermi's golden rule, the tunneling probability is

$$P_{i \to f} = \frac{2\pi}{\hbar} |M_e(k_{ez})|^2 (2\pi)^2 \delta\left(\boldsymbol{k}_{e\parallel} - \boldsymbol{k}'_{e\parallel}\right) \delta\left(\frac{\hbar^2 k_{ez}^2}{2m_e} - E_{ez} + eV_a\right) , \tag{4.8}$$

where M_e is the tunneling matrix element.

Since the Hamiltonian is time-reversible, to obtain the matrix element we consider the reverse process, where one electron inside the QW tunnels out to the n-doped GaAs layer. Assuming that all the electron states in the n-doped GaAs layer are empty, from (4.8) the electron tunneling-out rate can be expressed as follows:

$$\frac{1}{\tau_e} = \frac{m_e}{\hbar^3 k_{ez}} |M_e(k_{ez})|^2 . \tag{4.9}$$

Using the WKB approximation, we have calculated the electron tunneling-out rate in our structure to be

$$\frac{1}{\tau_e} \tag{4.10}$$

$$\simeq \frac{4\hbar k_{ez}}{w m_e} \exp\left(-\frac{4d}{3eV_a}\sqrt{\frac{2m_e}{\hbar^2}}[(V_e - E_{ez} + eV_a)^{3/2} - (V_e - E_{ez})^{3/2}]\right).$$

From (4.9) and (4.10), the tunneling matrix element is

$$|M_e(k_{ez})|^2 \simeq \frac{8\hbar^2(E_{ez} - eV_a)}{m_e w} \tag{4.11}$$

$$\times \exp(-\frac{4d}{3eV_a}\sqrt{\frac{2m_e}{\hbar^2}}[(V_e - E_{ez} + eV_a)^{3/2} - (V_e - E_{ez})^{3/2}]).$$

After the electrons tunnel resonantly into the QW, they either recombine with holes or tunnel back to the n-doped GaAs layer from the QW. From (4.10), we have estimated that the typical tunneling time constant for our structure is about 50 ps [116], which is much longer than the typical electron–hole recombination time constant ($\sim$ 10 ps) in the presence of many holes inside the QW. Therefore the electron population inside the QW is very small, and the tunneling-back current is negligible.

To obtain the resonant tunneling current density of electrons J_e, we integrate the tunneling rate over all the initial electron states in the n-doped GaAs layer and the final electron states in the QW, taking into account the Fermi–Dirac distribution of electrons in the initial states [117–119], and obtain

$$J_e = \frac{em_e^2|M_e|^2 k_B T}{2\pi\hbar^4\sqrt{2m_e(E_{ez} - eV_a)}} \ln(1 + e^{(E_{fe} - E_{ez} + eV_a)/k_B T}). \tag{4.12}$$

Using (4.11), we can find the asymptotic behavior of J_e at the upper and lower limits of the bias V_a. When V_a approaches its upper limit $V_{max} = E_{ez}/e$, the electron's velocity along the z direction $\hbar k_{ez}/m_e$ which satisfies the resonant tunneling condition is proportional to $(E_{ez} - eV_a)^{1/2}$, and thus goes to 0. Therefore J_e, which is proportional to k_{ez}, also approaches zero as $(V_{max} - V_a)^{1/2}$. When V_a approaches its lower limit $V_{min} = (E_{ez} - E_{fe})/e$ at $T = 0$ K, the number of electrons which can tunnel into the QW goes to zero as a function of $V_a - V_{min}$. Therefore J_e, which is proportional to the number of electrons that can tunnel, also approaches zero as $V_a - V_{min}$.

4.1.3 Exciton Formation by Hole-Assisted Resonant Tunneling of Electrons

Next, we calculate the current density of the hole-assisted resonant tunneling process of electrons. In this case, the initial state is a free electron in the n-doped GaAs layer and a subband hole in the QW, i.e. $|i\rangle = |\boldsymbol{k}_{e\|}, k_{ez}\rangle|\boldsymbol{k}_{h\|}, E_{hz}\rangle$, and the final state is a 2D exciton in the QW, i.e. $|f\rangle = |\boldsymbol{k}_{ex\|}, E_{ex}; E_{ez}, E_{hz}\rangle$. From Fermi's golden rule, the tunneling probability is

$$P_{i\to f} = \frac{2\pi}{\hbar} |M_{ex}(k_{ez})|^2 (2\pi)^2 \delta(\boldsymbol{k}_{e\|} + \boldsymbol{k}_{h\|} - \boldsymbol{k}_{ex\|})$$

$$\times\delta\left(\frac{\hbar^2 k_{ez}^2}{2m_e} + \frac{\hbar^2 k_r^2}{2\mu} - E_{ez} + E_{ex} + eV_a\right), \tag{4.13}$$

where M_{ex} is the tunneling matrix element.

Again, in order to obtain the matrix element M_{ex}, we consider the reverse process, i.e. an exciton in the QW dissolves into an electron and a hole and the electron tunnels from the QW to the n-doped GaAs layer. Assuming that all the electron states in the n-doped GaAs layer are empty, the tunneling-out rate can be expressed as follows:

$$\frac{1}{\tau_{ex}} = \int \frac{2\pi}{\hbar} |M_{ex}|^2 \, \delta \left(\frac{\hbar^2 k_{ez}^2}{2\mu} + \frac{\hbar^2 k_r^{\,2}}{2\mu} + eV_a - E_{ez} + E_{ex} \right)$$

$$\times \frac{1}{2\pi} dk_{ez} \frac{1}{(2\pi)^2} \, d\mathbf{k}_r \,. \tag{4.14}$$

Since the tunneling rate is independent of the initial exciton's center-of-mass transverse motion, we assume $k_{ex\parallel} = 0$, or, in other words, we move to the exciton's center-of-mass frame.

We have used the WKB method to estimate the tunneling-out rate $1/\tau_{ex}$. In a thin QW where the well thickness w is equal to or less than the exciton Bohr radius a_B, the 2D approximation for the exciton wavefunction inside the QW gives

$$\psi_i(\mathbf{r}_\parallel, z_e, z_h) = \left(\frac{8}{\pi a_B^{\,2}} \right)^{1/2} e^{-2r_\parallel / a_B} \, \psi_e(z_e) \psi_h(z_h) \,, \tag{4.15}$$

where $\mathbf{r}_\parallel$ is the electron–hole relative coordinate in the plane parallel to the QW plane; $\psi_e(z_e)$ and $\psi_h(z_h)$ are the confined electron and hole wavefunctions in the z direction. It should be pointed out that owing to the high hole density in the QW, the screening effect will modify the simple hydrogen-like 2D exciton wavefunction; this effect has been neglected in our first-order calculation.

For a high, thin barrier, as the electron tunnels through the barrier to its right-hand edge, the exciton wavefunction can be approximated as

$$\psi_i(\mathbf{r}_\parallel, z_e = d, z_h) = 2 \left(\frac{8}{\pi a_B^{\,2}} \right)^{1/2} e^{-2r_\parallel / a_B} \psi_e(z_e = 0) e^{-b/2} \psi_h(z_h) \,, \tag{4.16}$$

where

$$\kappa(z_e) = \sqrt{\frac{2m_e}{\hbar^2} \left(V_{eff} - E_{ez} + eV_a + \frac{\hbar^2 k_r^2}{2\mu} \frac{z_e}{d} \right)}, \tag{4.17}$$

$$b \equiv 2 \int_0^d \kappa(z_e) dz_e = \frac{4d\sqrt{2m_e}}{3\hbar(eV_a + \hbar^2 k_r^{\,2}/2\mu)}$$

$$\times [(V_{eff} - E_{ez} + eV_a + \hbar^2 k_r^{\,2}/2\mu)^{3/2} - (V_{eff} - E_{ez})^{3/2}], \tag{4.18}$$

where V_{eff} is the effective barrier height for the electron. Since the barrier height is increased by the exciton binding energy E_{ex} and decreased by the

Coulomb interaction between the electron in the barrier and the hole in the QW, the upper and lower bounds for V_{eff} are

$$V_e < V_{\text{eff}} < V_e + E_{\text{ex}} \tag{4.19}$$

From (4.19) and (4.18), we obtain the range of b. For the typical parameters of our structure, i.e. $V_e \sim 230$ meV, $E_{\text{ex}} \sim 10$ meV, and barrier width $d \sim 100$ Å, we found that $\Delta b/b < 0.3$. So the error in $e^{-b/2}$ is within 15%, no matter wether we use the upper or the lower bound for V_{eff}.

To match both the wavefunction and its derivative at the boundary between the barrier and the n-doped GaAs layer ($z_e = d$), we also take into account the reflected component of the wavefunction inside the barrier, which increases with z_e ($0 \leq z_e \leq d$). It is the reflected wave that results in the factor 2 on the right-hand side of (4.16).

After the exciton has dissolved into a free electron in the n-doped GaAs layer and a subband hole in the QW, the wavefunction becomes

$$\psi_f(\boldsymbol{r}_\|, z_e, z_h) = e^{i\boldsymbol{k}_r \cdot \boldsymbol{r}_\|} e^{ik_{ez}(z_e - d)} \psi_h(z) \,, \tag{4.20}$$

where $\boldsymbol{k}_r$ is the electron–hole relative transverse momentum.

We match the in-barrier wavefunction $\psi_i(\boldsymbol{r}_\|, z_e, z_h)$ with a linear combination of the outgoing wavefunction $\psi_f(\boldsymbol{r}_\|, z_e, z_h)$ at the boundary between the barrier and the n-doped GaAs layer:

$$\psi_i(\boldsymbol{r}_\|, z_e = d, z_h) = \int A(\boldsymbol{k}_r) \psi_f(\boldsymbol{r}_\|, z_e = d, z_h) dk_r \,. \tag{4.21}$$

From this, we obtain

$$A(\boldsymbol{k}_r) = 2\frac{8a_B}{\sqrt{2\pi}(4 + a_B^2 k_r^2)^{3/2}} \psi_e(z_e = 0) \, e^{-b/2} \,. \tag{4.22}$$

The tunneling rate is

$$\frac{1}{\tau_{\text{ex}}} = \int |A(\boldsymbol{k}_r)|^2 \frac{\hbar k_{ez}}{m_e} dk_r$$

$$= \int \frac{(8a_B)^2}{2\pi(4 + a_B^2 k_r^2)^3} \frac{4}{w} e^{-b} \frac{\hbar k_{ez}}{m_e} dk_r \,. \tag{4.23}$$

By comparing (4.14) and (4.23), we obtain the tunneling matrix element:

$$|M_{\text{ex}}(k_r)|^2 = \frac{2^8 (2\pi)^2 a_B^2 \hbar^4 k_{ez}^2}{m_e^2 w (4 + a_B^2 k_r^2)^3} e^{-b}. \tag{4.24}$$

Note that k_{ez} and k_r are related by the resonant-tunneling condition (4.5).

Before finishing the derivation of the tunneling current, we need to outline the classification of the excitons formed through tunneling.

Owing to the electron–hole exchange interaction, a noninteracting QW exciton should be described in terms of the z component of its total angular momentum, rather than the individual z components of the angular momentum of the electron and hole. However, the hole spin–orbit interaction combined with the QW confinement effect results in a mixing of LH and HH exciton states with different z components of exciton angular momentum [120] (compare with a similar mixing of HH and LH states for a QW hole when $k_{h\parallel} \neq 0$ [24]). But we may still classify excitonic states in terms of LH and HH excitons at small $k_{ex\parallel}$, where the mixing of LH and HH states is small.

The electron–hole exchange interaction (both the short-range exchange interaction and the longitudinal–transverse splitting [121]) splits both the QW LH and the QW HH exciton levels into four sublevels [122]. Of the total of eight sublevels, six correspond to superposition states of spin singlet and spin triplet states. Excitons in these states can relax to $k_{ex\parallel} = 0$ by phonon scattering and then radiatively decay. Excitons in the other two states (one for LH and one for HH excitons) are paraexcitons (i.e. their spin state is pure triplet). Paraexcitons are not radiatively active, so they are first converted into optically active excitons (by means of exciton–exciton and exciton–free-carrier interaction) and then radiatively decay. One-quarter of the total created LH and HH excitons are paraexcitons.

An alternative scenario for both types of excitons is for them to dissolve into a free electron and a subband hole, with the electron tunneling back to the n-doped GaAs layer from the QW. We have estimated that the typical exciton lifetime set by the back-tunneling process is about 50 ps, which is much longer than the typical exciton conversion, relaxation, and radiative-recombination lifetime ($\sim$ 10 ps). Therefore, the exciton population inside the QW is negligibly small, and the net exciton tunneling current is approximately given by the forward tunneling current.

To obtain the total tunneling current density J_{ex} into an exciton level, we integrate the tunneling rate over all possible combinations of initial and final states, taking into account the Fermi–Dirac distributions of electrons $f_e(\boldsymbol{k}_{e\parallel}, k_{ez})$ and holes $f_h(\boldsymbol{k}_{h\parallel}, E_{hz})$ in the initial states, and obtain

$$J_{ex} = \frac{m_e}{(2\pi)^4 \hbar^3} \int \frac{|M_{ex}(k_r)|^2}{k_{ez}} \, f_e(\boldsymbol{k}_{e\parallel}, k_{ez}) \, f_h(\boldsymbol{k}_{h\parallel}, E_{hz}) \, d\boldsymbol{k}_{e\parallel} \, d\boldsymbol{k}_{h\parallel} \, . \quad (4.25)$$

In (4.25), we can change the integration variables from $\boldsymbol{k}_{e\parallel}$ and $\boldsymbol{k}_{h\parallel}$ to $\boldsymbol{k}_{ex\parallel}$ and $\boldsymbol{k}_r$ by expressing $\boldsymbol{k}_{e\parallel}$ and $\boldsymbol{k}_{h\parallel}$ in terms of $\boldsymbol{k}_{ex\parallel}$ and $\boldsymbol{k}_r$. Since the integrand depends only on the magnitudes of $k_{ex\parallel}$ and k_r and the angle θ between $\boldsymbol{k}_{ex\parallel}$ and $\boldsymbol{k}_r$, we can simplify (4.25) to

$$J_{ex} = \int k_{ex\parallel} \, dk_{ex\parallel} \int k_r \, dk_r \quad (4.26)$$

$$\times \int d\theta \, \frac{m_e}{8\pi^3 \hbar^3} \frac{|M_{ex}(k_r)|^2}{k_{ez}} \, f_e(k_{ex\parallel}, k_r, \theta) \, f_h(k_{ex\parallel}, k_r, \theta) \, . \quad (4.27)$$

Before we calculate J_{ex} numerically, let us estimate the peak value of J_{ex} as compared with that of J_e. Since $k_r \sim 1/a_B$, from (4.24), we have $|M_{ex}|^2 \sim 2\pi a_B^2 |M_e|^2$. From (4.27), we obtain

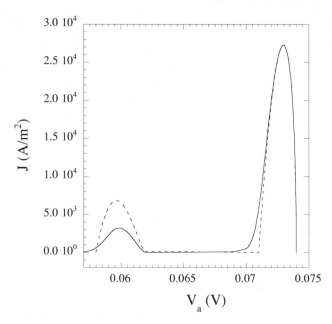

Fig. 4.2. Numerically calculated tunneling-current density as a function of the bias voltage at $T = 0$ K (*dashed line*) and $T = 4$ K (*solid line*)

$$J_{\text{ex}} \sim 2\,\pi a_{\text{B}}^2\,N_{\text{h}}\,J_e\,,\tag{4.28}$$

where N_{h} is the 2D hole density inside the QW. For $\pi a_{\text{B}}^2\,N_{\text{h}} \sim 0.1$, the ratio of J_{ex} over J_e is about 0.2. Hence J_{ex} is comparable in magnitude to J_e.

Figure 4.2 shows the calculated tunneling-current density as a function of the effective bias V_{a} for $T = 0$ K (dashed line) and $T = 4$ K (solid line). The QW considered here is 50 Å wide, and the $\text{Al}_{0.3}\text{Ga}_{0.7}\text{As}$ barrier is 100 Å wide. The 1s HH exciton binding energy E_{ex} in the QW is about 12 meV [34]. The electron Fermi level in the n-doped GaAs layer is 3 meV above the conduction band edge, and the hole Fermi level in the QW is 1 meV below the first hole subband edge in the QW, corresponding to $\pi a_{\text{B}}^2\,N_{\text{h}} \sim 0.08$. The first current peak corresponds to resonant tunneling into the 1s HH exciton level below the first electron subband. The second current peak represents the normal resonant tunneling of electrons into the first electron subband in the QW. We have neglected the fine structure of the exciton tunneling current due to the exciton level splitting, since the splitting is quite small. Figure 4.2 shows that at finite temperature, J_{ex} has a tail in the low-bias side, and also that its peak value decreases.

We can estimate the asymptotic behavior of J_{ex} at the upper and lower limits of the bias V_{a} using (4.25). When V_{a} approaches its upper limit $V_{\text{max}} = (E_{ez} - E_{\text{ex}})/e$, the z component velocity $\hbar k_{ez}/m_e$ of the electrons which can tunnel in approaches 0, since it is proportional to $(V_{\text{max}} - V_{\text{a}})^{1/2}$. Since

$k_r \leq \sqrt{2\mu(E_{ez} - E_{ex} - eV_a)}/\hbar$, k_r also goes to 0. This means electrons can only combine with holes with the same lateral velocity to form excitons. Both restrictions lead to the asymptotic behavior $J_{ex} \sim (V_{max} - V_a)^{3/2}$.

On the other hand, when V_a approaches its lower limit (4.7) at $T = 0$ K, the numbers of electrons and holes which can combine to form excitons both approach zero as $V_a - V_{min}$. In addition, the total transverse momentum of an electron–hole pair also goes to zero because energy conservation only allows a small $k_{ex\|}$. Therefore,

$$J_{ex} \sim \begin{cases} (V_a - V_{min})^2 , & \text{if } k_{fe} > k_{fh} , \\ (V_a - V_{min})^{5/2} , & \text{if } k_{fe} < k_{fh} . \end{cases} \qquad (4.29)$$

The extra factor $(V_a - V_{min})^{1/2}$ for $k_{fe} < k_{fh}$ appears because the z component velocity of the tunneling electrons is proportional to $(V_{max} - V_a)^{1/2}$. Therefore the asymptotic behavior of J_{ex} is different from that of J_e owing to its two-particle nature.

4.1.4 Selective Creation of Excitons by Tunneling

This hole-assisted resonant tunneling process of electrons can create excitons not only directly, but also selectively.

For comparison, let us start with the one-particle tunneling process, and consider an electron tunneling into an electron subband level in the QW. Because the electron's transverse momentum $k_{e\|}$ is conserved during the tunneling, the electron's transverse kinetic energy $\hbar^2 k_{e\|}^2 / 2m_e$ is also conserved and cancels out in the energy conservation equation. Therefore the resonant-tunneling condition (4.6) has nothing to do with the electron's transverse momentum $k_{e\|}$. This means the initial $k_{e\|}$ distribution of the electrons outside the QW is equal to the final $k_{e\|}$ distribution of the electrons in the QW after the tunneling. Since the initial electrons are in a thermal bath, the final $k_{e\|}$ distribution of the electrons in the QW remains a thermal distribution.

In the case of hole-assisted resonant tunneling of electrons into an exciton state, although the sum of the initial transverse momenta of the electron and hole is equal to the final exciton's transverse momentum (4.1), the sum of the initial transverse kinetic energies of the electron and hole can be larger than the final exciton's kinetic energy (4.2). In the low-bias regime where $eV_a < E_{ez} - E_{ex} - E_{fe}$, even though the electron's perpendicular energy $\hbar^2 k_{ez}^2 / 2m_e$ is not enough to reach the QW exciton level $E_{ez} - E_{ex} - eV_a$, the electron can still tunnel into the QW and combine with a hole with *opposite* transverse momentum to form an exciton with almost zero transverse momentum, so that the electron–hole's excess transverse kinetic energy $\hbar^2 k_r^2 / 2\mu$ adds to $\hbar^2 k_{ez}^2 / 2m_e$ to reach $E_{ex} + E_{ez} - eV_a$. In this sense, this tunneling process is a *hole-assisted* resonant tunneling process that directly creates excitons in the QW. It is a new type of assisted tunneling (phonon-assisted tunneling is the most prominent representative of assisted tunneling [125, 126]). More

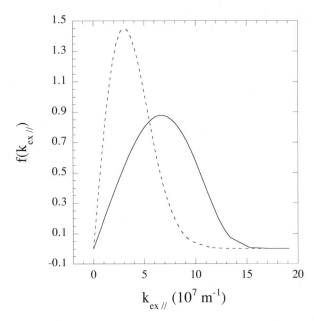

Fig. 4.3. Calculated $k_\parallel$ distribution of QW excitons formed by hole-assisted resonant tunneling of electrons at $T = 4$ K under two bias voltages: $V_a = 0.054$ V (*dashed line*) and $V_a = 0.061$ V (*solid line*)

importantly, excitons formed under such low bias have a very small transverse momentum. In other words, this hole-assisted resonant tunneling process of electrons can selectively create *cold* excitons from the thermal bath. At $T = 0$ K, the range of $k_{ex\parallel}$ under a bias V_a is

$$0 \leq k_{ex\parallel} \leq \frac{\sqrt{2Me(V_a - V_{min})}}{\hbar} \ .$$

Therefore, as the bias voltage V_a approaches the minimum bias V_{min}, most of the excitons created through tunneling have $\boldsymbol{k}_{ex\parallel} \simeq 0$.

For finite temperature, we can easily obtain from (4.27) the transverse-k-vector ($k_{ex\parallel}$) distribution $f(k_{ex\parallel})$ of the excitons formed by hole-assisted resonant tunneling of electrons before thermalization occurs:

$$f(k_{ex\parallel}) \propto k_{ex\parallel} \int dk_r \frac{k_r \, |M_{ex}(k_r)|^2}{k_{ez}} \int d\theta \, f_e(\boldsymbol{k}_{ex\parallel}, k_r, \theta) \, f_h(\boldsymbol{k}_{ex\parallel}, E_r, \theta) \ . \tag{4.30}$$

Figure 4.3 shows the exciton $k_{ex\parallel}$ distribution $f(k_{ex\parallel})$ under two different biases at $T = 4$ K. We can see that at lower bias (dashed line), more excitons are formed with smaller transverse momentum. As we know, excitons with small $\Delta k_{ex\parallel}$ have a large spatial coherence, leading to a superradiative decay [38]. For example, excitons with $k_{ex\parallel} \leq 3 \times 10^7$ m^{-1} have a spatial-coherence

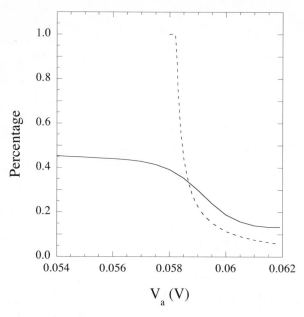

Fig. 4.4. Percentage of excitons created by hole-assisted resonant tunneling of electrons with $k_{\mathrm{ex}\,\|} < 3 \times 10^7$ m^{-1} at $T = 0$ K (*dashed line*) and $T = 4$ K (*solid line*)

area larger than 300 Å in radius, and their radiative decay rate is one order of magnitude larger than that of excitons without spatial coherence [124]. Figure 4.4 shows the percentage of excitons created with a spatial-coherence area larger than 300 Å in radius. At $T = 0$ K, the percentage increases to 100% when the bias is low enough, while at $T = 4$ K the percentage saturates at about 50% at low bias.

4.2 Resonant Tunneling into QW Exciton States – Experiment

4.2.1 *I–V* Measurement

Tunneling structures, as shown in Fig. 4.1 were grown by MBE on n$^+$ ⟨100⟩ GaAs substrates. In the first sample grown, the GaAs QW was 100 Å thick, and the Al$_{0.3}$Ga$_{0.7}$As barrier was 75 Å thick. The p-doped Al$_{0.3}$Ga$_{0.7}$As layer was 3000 Å thick, and doped at 1×10^{18} cm^{-3}. The n-doped GaAs layer was 3000 Å thick, and doped at 4×10^{16} cm^{-3}. We fabricated micropost structures with lateral dimensions between 10 and 100 μm using wet etching. The sample was cooled to 4.2 K. Its *I–V* characteristics were measured by an HP 4155A semiconductor parameter analyzer in the constant-voltage operation mode.

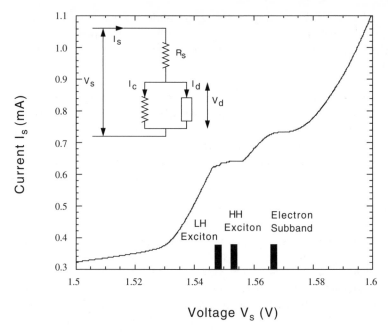

Fig. 4.5. Measured current I_s as a function of the applied voltage V_s for the first sample (see text), at 4.2 K. The *inset* shows the equivalent circuit of the $I\text{--}V$ measurement

The inset of Fig. 4.5 shows the equivalent measurement circuit. The measured current I_s is the sum of the tunneling current I_d and the leakage current I_c. The leakage current I_c is due to surface recombination, thermionic emission of electrons into the QW, and thermionic emission and tunneling of holes from the QW to the n-doped GaAs layer. The measured voltage $V_s = V_d + I_s R_s$, where V_d is the voltage drop across the tunnel junction and R_s is the resistance in series with the tunnel junction, including the resistance of the ohmic contact, the bulk layers, and the substrate.

As shown in Fig. 4.5, the measured $I_s - V_s$ curve of the first sample clearly shows two resonant-tunneling peaks very close to each other. The series resistance R_s was measured to be about 3 Ω. By subtracting the voltage drop $I_s R_s$ across the series resistance from the measured V_s, we found the actual separation in junction voltage (V_d) between these two peaks to be about 14 meV. According to theoretical calculations [34] and photoluminescence excitation (PLE) experiments [127–129], the LH exciton binding energy in a 100 Å thick GaAs/Al$_{0.3}$Ga$_{0.7}$As QW is about 13 meV, and the HH exciton binding energy is about 10 meV. Therefore we believe that the first current peak corresponds to the resonant tunneling of electrons into the exciton levels below the first electron–hole subband in the QW, and the second current peak corresponds to the resonant tunneling of electrons into the first electron subband level in the QW. Owing to the large inhomogeneous broadening of the QW energy levels in this sample, the tunneling peaks are too broad to separate the tunneling peak into the LH exciton level from that into the HH exciton level.

In order to reduce the width of the tunneling peaks, we grew a second sample with a narrower QW (50 Å) to increase the separation between the QW energy levels. We also introduced growth interruption at the QW interfaces to reduce the QW thickness fluctuations. To further narrow each tunneling peak, we decreased the doping level of the n-doped GaAs layer to 8×10^{15} cm^{-3}. The Al$_{0.3}$Ga$_{0.7}$As barrier was 110 Å thick. Unfortunately, in this sample the series resistance R_s is increased to 15 Ω, owing to the low doping of the n-doped GaAs layer. Since the series resistance R_s is now larger than the absolute value of the negative differential resistance of the tunnel junction, a resonant-tunneling peak in the I_s–V_d characteristic of the tunnel junction corresponds to a hysteresis loop in the measured I_s–V_s curve (Fig. 4.6a). This is because the measured I_s corresponds to the intersection point of the load line $I_s = (V_s - V_d)/R_s$ and the I_s–V_d curve of the tunnel junction. As V_s increases, I_s will increase until it reaches the resonant-tunneling peak (point 1 in Fig. 4.6a), where I_s jumps downward from point 1 to point 2, and then increases again. The sudden drop of the current I_s results in a negative spike in the differential conductance dI_s/dV_s. On the other hand, when V_s decreases, I_s decreases until it reaches the valley (point 3), where it jumps upward from point 3 to point 4, and then decreases again.

Figure 4.6b shows the measured differential conductance dI_s/dV_s as a function of the bias voltage V_s when V_s is swept upward and downward. The three negative spikes in dI_s/dV_s as V_s is swept upward and the three slightly down-shifted spikes in dI_s/dV_s as V_s is swept downward correspond to three hysteresis loops, which indicate the existence of three resonant-tunneling current peaks. After subtracting $I_s R_s$ from V_s, we found that the actual junction voltage separation between the first and third tunneling peak is close to the value of the 1s LH exciton binding energy (∼ 16 meV) in a 50 Å GaAs QW, and the separation in V_d between the second and third tunneling peak is close to the value of the 1s HH exciton binding energy (∼ 12 meV) in a 50 Å GaAs QW [34, 127]. Therefore this suggests that the three tunneling peaks correspond to the resonant tunneling of electrons into the LH exciton level, the HH exciton level, and the first electron subband level in the QW.

4.2.2 L–I Measurement

We also measured the light emitted by the QW excitons created through the hole-assisted resonant tunneling process of electrons. In order to measure the light from the QW, we fabricated some tunnel diodes whose top metal contacts had some open area so that the photons emitted by the QW excitons could partially escape from the top. A photodetector with an extremely linear response to weak light was placed in front of the device to measure the light

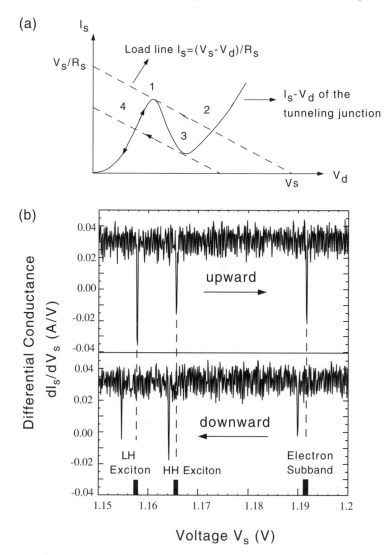

Fig. 4.6. (a) A resonant-tunneling peak in the I_s–V_d curve results in a hysteresis loop in the I_s–V_s curve when the series resistance R_s is larger than the absolute value of the negative differential resistance of the tunnel junction. (b) The measured differential conductance dI_s/dV_s as a function of the bias voltage V_s for the second sample when V_s is swept upward and downward

emitted into an 18° half-angle cone in the air (which corresponds to about 5° in the GaAs QW).

Figure 4.7a shows the measured current I_s of such a tunnel diode as a function of the bias voltage V_s. The three current peaks in the I_s–V_s curve re-

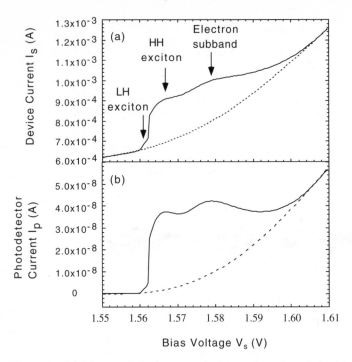

Fig. 4.7. (a) Measured device current I_s as a function of the bias voltage V_s. (b) Measured photodetector current I_p as a function of the bias voltage V_s

sult from the resonant tunneling of electrons into the 1s LH exciton state, the 1s HH exciton state, and the electron subband level. Owing to the valence-band splitting in the QW, there are many more heavy holes than light holes in the QW. Hence the LH exciton tunneling peak is much smaller than the HH exciton tunneling peak. Moreover, acoustic-phonon-assisted tunneling results in long tails of the tunneling peaks on the higher-bias side. Figure 4.7b shows the light-induced photodetector current I_p as a function of the bias voltage V_s applied to the device. In the I_p–V_s curve, there are three light emission peaks corresponding, to the three resonant tunneling current peaks in the I_s–V_s curve. However, the QW excitons created through thermionic emission of electrons from the n-doped GaAs layer to the QW and subsequent relaxation can also emit photons, leading to the background current in I_p. We curve-fitted the background current in the neighborhood of the tunneling peaks (see the dashed line in Fig. 4.7b) and subtracted it from I_p to obtain the photodetector current I_e which is induced only by the emission from the excitons created through resonant tunneling. In the same way, we also removed from I_s the background current due to the thermionic emission of electrons from the n-doped GaAs layer into the QW and to tunneling and

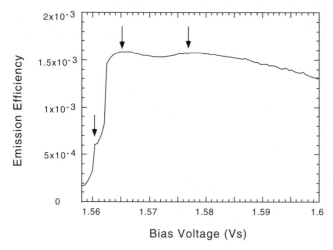

Fig. 4.8. Measured exciton emission efficiency into the normal direction within a 5° half-angle as a function of the bias voltage V_s

thermionic emission of holes from the QW to the n-doped GaAs layer, and thus obtained the pure electron tunneling current I_d.

After taking into account the efficiency of the photodetector ($\sim 65\%$), the sample's surface transmissivity ($\sim 70\%$), and the percentage of open area in the top contact ($\sim 25\%$), we converted the ratio of I_e to I_d into the exciton emission efficiency into the normal direction within a 5° half-angle (see Fig. 4.8). As is well known, the ratio of the HH exciton emission efficiency to the LH exciton emission efficiency into the normal direction (i.e., with their in-plane k vector $k_\parallel = 0$) is equal to the ratio of their in-plane oscillator strengths, which is 3:1 [23]. In Fig. 4.8, the ratio of the HH exciton emission efficiency to the LH exciton emission efficiency is about 2.7, which is slightly smaller than 3. This discrepancy is caused by the fact that we collected light from excitons with $k_\parallel < 2.4 \times 10^6$ m^{-1}, owing to the finite light collection angle (5° in the GaAs). For excitons with nonzero $k_\parallel$, the valence-band mixing changes the ratio of the in-plane oscillator strengths of the HH exciton to the LH exciton from 3 toward 2 [23].

More quantitatively, we have calculated the emission efficiency of both HH excitons and LH excitons into the normal direction within a 5° half-angle, taking into account the different dependences of the emission rate on the emission angle for transverse, longitudinal, and Z (dipole moment perpendicular to the QW plane) excitonic modes [23]. Owing to the rapid thermalization process, we assume that the $k_\parallel$ distribution of the excitons created through the resonant tunneling process of electrons has reached a thermal distribution before photon emission. With the parameters used in our experiment, the calculated emission efficiency for HH excitons is about 3.41×10^{-3}, and the calculated emission efficiency for LH excitons is about 1.26×10^{-3}. The measured efficiencies are smaller than these theoretical values by a factor of 2. This may be due to the creation of optically inactive excitons with spin 1 and their subsequent nonradiative recombination before spin relaxation. However, the theoretical and experimental ratios of HH exciton emission efficiency to LH exciton emission efficiency are both 2.7.

In addition, since there are many more heavy holes than light holes in the QW, most of the electrons, after tunneling into the electron subband level of the QW, form HH excitons with large $k_\parallel$ and then thermalize to $k_\parallel \simeq 0$ before photon emission. Therefore the emission efficiency of the electrons tunneling into the electron subband level is close to the emission efficiency of the electrons tunneling into the HH exciton level.

4.2.3 Tunneling Spectroscopy for QW Excitons

It turns out that this hole-assisted resonant tunneling process of electrons into a QW exciton level also provides a new technique for the tunneling spectroscopy of QW excitons.

The exciton binding energy E_{ex} determines how robust, or how useful, the exciton is in practical applications [130]. Unfortunately, the value of E_{ex} cannot be extracted from the usual PL or PLE spectra, where the continuum edge $E_c - E_v$ is not observable. The conventional way of measuring E_{ex} is optical-absorption measurement [131]. It turns out hole-assisted resonant tunneling of electrons into the QW exciton state can also be used to directly measure the QW exciton binding energy, because the bias voltage separation between the resonant-tunneling-current peak of electrons into the exciton level and the resonant-tunneling-current peak of electrons into the electron subband level is equal to the exciton binding energy.

Figure 4.9 illustrates the conceptual difference between this form of tunneling spectroscopy and conventional optical absorption or emission spectroscopy. In the emission spectrum, the frequency of an exciton peak ($E_c - E_v - E_{ex}$) is determined not only by the exciton binding energy E_{ex}, but also by the electron subband energy E_c and the hole subband energy E_v. Therefore the 1s LH exciton peak appears at an energy *higher* than that of the 1s HH exciton peak in the emission spectrum, as shown in Fig. 4.9c. On the other hand, in tunneling spectroscopy, the voltage separation between the resonant-tunneling-current peak of electrons into an exciton level and that into the electron subband level is determined only by the exciton binding energy. Therefore the resonant-tunneling-current peak of electrons into the 1s LH exciton state appears at a bias voltage *lower* than that into the 1s HH exciton state (see Fig. 4.9b).

One advantage of this tunneling spectroscopy as compared with conventional optical absorption or emission spectroscopy is that it can measure the binding energy of *optically inactive* excitons. Since an optically inactive exciton can neither absorb nor emit a photon, it cannot be observed in absorption or emission spectra. However, an optically inactive exciton can be created in

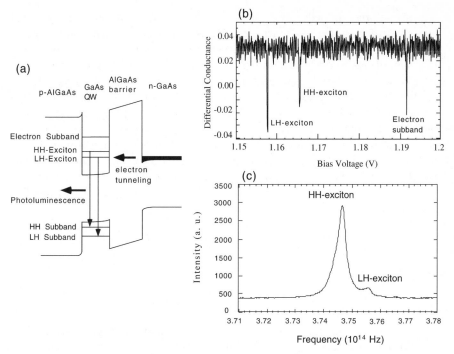

Fig. 4.9. (a) Band structure, illustrating the difference between tunneling spectroscopy and emission spectroscopy. (b) Measured differential conductance versus the bias voltage for the tunneling diode described in the text. (c) Measured photoluminescence spectrum of a 100 Å GaAs layer at 4.2 K

a QW through the hole-assisted resonant tunneling of electrons at an appropriate bias. In our present samples, the inhomogeneous broadening of the QW energy levels is still too large to separate the resonant-tunneling-current peak of electrons into the optically inactive exciton state from that into the optically active exciton state. Nonetheless, in principle, this tunneling spectroscopy can measure the binding energy of both optically active excitons and optically inactive excitons, and hence it can measure the singlet–triplet splitting of excitons, which could not be measured before.

4.3 Stimulated Resonant Tunneling

In the previous section, only the spontaneous tunneling process was analyzed. We may expect stimulated tunneling into exciton states having large occupation numbers to occur, as excitons behave as bosons at low density. For example, large populations of optically active excitons can be injected with an external optical pump. In this case, we may expect a direct dependence of the tunneling current on the energy flux of the external optical source.

The stimulated tunneling process is relevant to the creation of coherent exciton or exciton polariton matter waves, as explained later. In the following, we analyze and calculate the stimulated tunneling rates as a function of the bias conditions and the exciton density, and we show that under suitable conditions, stimulated exciton creation can be realized and detected.

We use nondimensional units for length and energy, scaled with the exciton Bohr radius and the exciton binding energy, respectively. At exciton densities N_{ex} much lower than the 2D exciton packing density $N_M = 1/(4\pi)$, the deviation of excitonic statistics from that of bosons is negligible. Therefore, the stimulated tunneling rate in the dilute regime is the product of the spontaneous tunneling rate and the exciton filling factor. The total tunneling current into excitonic states also includes a negative, backward tunneling process:

$$J = \frac{em_e}{(2\pi)^4\hbar^3} \int d\mathbf{k}_{e\|}\, d\mathbf{k}_{h\|} \frac{|M_{ex}(k_r)|^2}{k_{ez}} [(1 + n_{ex})f_e f_h - n_{ex}(1 - f_e)(1 - f_h)]$$

$$= J_{sp} + J_{ex}, \tag{4.31}$$

$$J_{ex} = \frac{em_e}{(2\pi)^4\hbar^3} \int d\mathbf{k}_{e\|}\, d\mathbf{k}_{h\|} \frac{|M_{ex}(k_r)|^2}{k_{ez}} n_{ex} [f_e + f_h - 1] \tag{4.32}$$

where e is the elementary charge, and $f_e(\mathbf{k}_{e\|}, k_{ez})$, $f_h(\mathbf{k}_{h\|})$, and $n_{ex}(\mathbf{k}_{ex})$ are the electron, hole, and exciton filling factors. k_{ez} is the momentum of the tunneling electron in the growth direction, and is determined by energy conservation: $(k_{e\|}^2 + k_{ez}^2)/2m_e + k_{h\|}^2/2m_h = -E_{ex} + k_{ex}^2/2m_{ex} + V$. We have subdivided J into two parts, a spontaneous part J_{sp}, which has been already calculated in (4.25), and a part proportional to the exciton filling J_{ex}, given by the difference between the stimulated tunneling current – corresponding to the $n_{ex}f_e f_h$ term in the square brackets of (4.31) – and the backward tunneling current – corresponding to the second term in the square brackets. The backward current in (4.31) represents a process of ionization of the excitons, which break up into bulk conduction electrons and QW holes. The backward current is therefore present even when $V < V_{min}$ (see (4.7)), when excitons are present in the QW. At larger biases $V > V_{min}$, the forward current eventually becomes larger than the backward one. In particular, if the momentum distribution of excitons is strongly centered around $k_{ex} = 0$, then the sign of J_{ex} is that of $V - V_{min}$. We plot in Fig. 4.10 the total tunneling current (solid line) and its spontaneous part (dashed line) as functions of $V - V_{min}$, where the electron and hole Fermi energies are $E_{hF} = 0.05$ and $E_{eF} = 0.15$, respectively, the temperature of the electrons and holes is $T = 0.01$ ($T \approx 1$ K for GaAs systems), $m_{ex} \approx 2.1$ as for GaAs, and $N_{ex} = 0.025$. We also suppose that the excitons have small kinetic energies $E_{kin} \ll T$, which is the case for optically injected excitons. We clearly see stimulation of the tunneling current at positive $V - V_{min}$; the ratio of the stimulated rate to the spontaneous rate is a maximum when $V - V_{min} \sim T$, as shown in the inset of Fig. 4.10. This behavior of J_{ex}/J_{sp} is due to the different rates of increase of J_{ex} and J_{sp} with

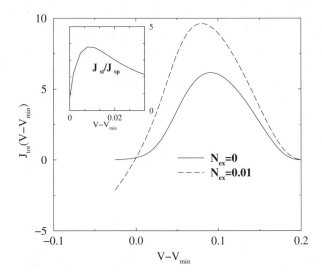

Fig. 4.10. The spontaneous (*solid line*) and total (*dashed line*) tunneling currents as a function of the bias voltage V, measured from the threshold $V_{\min}$. Energies are in units of E_{ex}. The *inset* shows the ratio of the stimulated to the spontaneous tunneling current versus $V - V_{\min}$

$V - V_{\min}$. In fact, $J_{\mathrm{ex}} \propto V - V_{\min}$, whereas we can see from Fig. 4.10 that J_{sp} (solid curve) is superlinear with respect to the bias. This superlinear behavior is related to the enlarged space of final tunneling states N_{states} available at larger bias. For $V - V_{\min} \sim T$, $N_{\mathrm{states}} \sim m_{\mathrm{ex}}T/(2\pi)$, and the ratio of the stimulated to the spontaneous tunneling current is $\sim N_{\mathrm{ex}}/N_{\mathrm{states}}$. Thus, this argument shows that stimulated tunneling can dominate over spontaneous tunneling in low-temperature conditions.

4.3.1 Saturation of the Stimulated Current

When the exciton density approaches the packing density, it is no longer possible to consider excitons as independent bosons, because their constituents are fermions. This results in a partial blocking (saturation) of electron tunneling into excitons (as well as into the QW subbands) because of the Pauli exclusion principle. At relatively low exciton densities, the exchange of exciton constituents with another exciton constituent or a single carrier can be described as an effective interaction between them. However, at higher densities, this interacting-boson description loses validity, and the ground state of the electron–hole system is better described as a coherent state of electron–hole pairs in momentum space [132], at least at zero temperature. At a finite exciton temperature T_{ex}, the condensate is replaced by a so-called "hunting condensate" [133], where the width of the momentum distribution of a macroscopic number of excitons is of the order of the the exciton thermal momentum $\sim \sqrt{m_{\mathrm{ex}} T_{\mathrm{ex}}}$. Therefore, electron tunneling is also stimulated at low temperatures $T_{\mathrm{ex}} \ll 1$, as long as $V - V_{\mathrm{min}} \gg T_{\mathrm{ex}}$. In this case, the width of the momentum distribution of the hunting condensate is clearly irrelevant in the calculation of the tunneling current, and we may assume $T_{\mathrm{ex}} = 0$ in the calculations. Moreover, the effect of finite electron and hole temperatures on the tunneling current is negligible if, likewise, $V - V_{\mathrm{min}} \gg T_{\mathrm{e,h}}$. In the following we concentrate on a simplified model with spinless carriers in an isotropic (in-plane) parabolic subband as in [132]. The generalization to several spin states per carrier momentum state is straightforward.

In the pairing approximation, the ground state is described by a BCS-like wavefunction [132, 134] given by

$$|\mathrm{gnd}\rangle = \prod_k (u_k + v_k a_k^\dagger b_{-k}^\dagger)|\mathrm{vac}\rangle , \qquad (4.33)$$

where $a_k^\dagger$ and $b_k^\dagger$ are the creation operators of an electron and a hole, respectively, in the lowest QW subbands with in-plane momentum k. The coefficients u_k and v_k, subject to the normalization condition $|u_k|^2 + |v_k|^2 = 1$, are found by minimizing the energy measured from the chemical potential $\Omega = \langle H - \mu N \rangle$, where $\mu = \mu_{\mathrm{e}} + \mu_{\mathrm{h}}$ is the chemical potential of the electron–hole pairs. The coefficients may be written in the following form [132]:

$$u_k v_k = \frac{\Delta_k}{2E_k} , \quad |v_k|^2 = \frac{1 - \zeta_k/E_k}{2} , \qquad (4.34)$$

where

$$\zeta_k = \epsilon_k - \mu - 2 \sum_{k'} V_{k-k'}|v_k|^2 = \zeta_{\mathrm{e}k} + \zeta_{\mathrm{h}k}$$

is the pair energy including the Fock terms measured from μ, $E_k = \sqrt{\zeta_k^2 + \Delta_k^2}$ is the energy required to extract an electron–hole pair from the condensate, $V_{k-k'}$ is the Coulomb interaction, which may be considered unscreened when $T \ll 1$ and $N_{\mathrm{h}} \ll 1/2\pi$, so that the screening due to unbound carriers is negligible. The condensation parameter is $\Delta_k = 2 \sum_{k'} V_{k-k'} u_k v_k$. In order to take into account the presence of excess holes in the QW, we construct the wavefunction of a state with a single hole-like excitation of momentum k as

$$|h_k\rangle = b_k^\dagger \prod_{k' \neq -k} (u_{k'} + v_{k'} a_{k'}^\dagger b_{-k'}^\dagger)|\mathrm{vac}\rangle , \qquad (4.35)$$

which has an excitation energy $E_{\mathrm{h}k} = (\zeta_{\mathrm{h}k} - \zeta_{\mathrm{e}k} + E_k)/2$.

When we calculate the tunneling current, we also assume that $N_{\mathrm{h}} \ll N_{\mathrm{M}}$, i.e. the hole gas is also dilute and its effect on the exciton system is negligible. The stimulated tunneling current is then given by

$$J_{\mathrm{ex}} = \frac{em_{\mathrm{e}}}{(2\pi)^4 \hbar^3} \int d\mathbf{k}_{\mathrm{e}\|} d\mathbf{k}_{\mathrm{h}\|} \frac{|M_{\mathrm{e}}|^2}{k_{\mathrm{e}z}} |v_k|^2 (f_{\mathrm{e}} + f_{\mathrm{h}} - 1), \qquad (4.36)$$

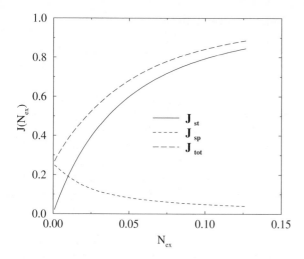

Fig. 4.11. Stimulated (*solid line*), spontaneous (*short-dashed line*) and total (*long-dashed line*) tunneling currents as functions of the exciton density N_{ex}, in units of a_{ex}^{-2}

which reduces to the J_{ex} of (4.32) when most excitons have small momenta $k_{ex} \approx 0$ and the exciton density is so low that $|v_k|^2 \approx 8\pi/(1 + k^2)^3$ [132]. The dependence of J_{ex} on the exciton density simplifies when the Fermi momentum of the electrons $k_{eF} \gg k_{hF}$, so that we may neglect variations in k_{ez}, and when $V - V_{min}$ is so large that *all* the holes assist the electron in tunneling:

$$k_{eF} \gg k_{hF} \text{ and } V - V_{min} > E_{hk=k_{hF}} - E_{hk=0}. \tag{4.37}$$

Under these conditions the stimulated tunneling rate is proportional to N_h and does not depend on the details of the hole dispersion curve:

$$J_{ex} \propto \int d\mathbf{k} |v_{\mathbf{k}}|^2 f_h(\mathbf{k}). \tag{4.38}$$

In particular, the stimulated tunneling current is just proportional to the exciton occupation probability $|v_{k=0}^2|$ for small dopings. We plot this probability as a function of the exciton (or electron–hole pair) density N_{ex} in Fig. 4.11 (solid line). This shows a sublinear dependence of the stimulated tunneling current on the exciton concentration, as expected from the Pauli exclusion principle, i.e. the blocking of the final states for tunneling at large exciton densities. We also notice that the stimulated current saturates. This is evident from (4.38), if we take $|v_k|^2 \sim 1$. Thus, we have normalized all the tunneling currents in Fig. 4.11 to the saturated value J_0.

4.3.2 Saturation of the Spontaneous Tunneling Current

In order to predict the *total* tunneling current, we also need to know the spontaneous tunneling current (J_{sp}). If the excitons were noninteracting bosons, J_{sp} would not depend on the exciton density at all. Instead, because the exciton constituents are fermions, we expect J_{sp} to be modified by the exclusion principle at large exciton densities. We shall restrict our consideration to small bias, where the effects of stimulated tunneling on the total current are largest, as explained before. J_{sp} arises from by tunneling into the excited states of the condensate. The low-energy excited states are collective modes, and in the pairing approximation these are asymptotically close to phase modes (Goldstone modes) (see e.g. [135]). This asymptotic approximation then holds for large exciton densities. For small exciton densities, low-energy excited states in the pairing approximation are not well approximated by phase modes, and are cumbersome to use. It is then easier to resort to an interacting-boson approximation, where the low-energy excited states are calculated directly through a Bogolyubov transformation, also taking into account vertex corrections at the ladder level.

In order to calculate the spontaneous tunneling rate, we need to express the collective modes of the system in terms of single-particle creation and annihilation operators [136,137]. In the pairing approximation, these modes can be constructed by transforming the ground state with an operator $\phi(\rho)$ which slowly rotates the condensate phase in space:

$$\exp\left(\mathrm{i}\int \mathrm{d}\rho\,\phi(\rho)\frac{[m_e a^\dagger(\rho)a(\rho) + m_h b^\dagger(\rho)b(\rho)]}{m_e + m_h}\right). \tag{4.39}$$

The Hamiltonian can then be written in terms of the phase operators $\phi(\boldsymbol{\rho})$ and their conjugates $N_{ex}(\rho) = -\mathrm{i}\partial/\partial\phi(\boldsymbol{\rho})$, which represent the *local* exciton concentration:

$$\hat{H} = -\frac{1}{2}\frac{\partial\mu}{\partial N_{ex}}\sum_q \frac{\partial}{\partial\phi_q}\frac{\partial}{\partial\phi_{-q}} + \frac{N_{ex}}{2m_{ex}}\sum_q \phi_q\phi_{-q}, \tag{4.40}$$

where μ is the exciton chemical potential, N_{ex} is the *average* exciton concentration, and ϕ_q are the Fourier components of $\phi(\rho)$. The Hamiltonian (4.40) shows acoustic-phonon eigenmodes with energy $\omega_q \equiv c_s q = \sqrt{(\partial\mu/\partial N_{ex})N_{ex}/m_{ex}}\,q$ for sufficiently small $q \ll c_s m_{ex}$, or $\omega_q \gg q^2/(2m_{ex})$. In spontaneous tunneling, an electron tunnels and combines with a hole-like single-particle excitation, described by (4.35), to produce a final state with one collective excitation. In the following, we also strengthen the second condition in (4.37) to $V - V_{min} \gg E_{hk=k_{hF}} - E_{hk=0}$, in order to completely neglect any hole mass renormalization effects [138]. For $N_{ex} < 1$, if $V - V_{min} \ll 6N_{ex}/|\log N_{ex}|$, where $6N_{ex}/|\log N_{ex}|$ is the energy of interaction between the excitons in the pairing approximation, the collective excitations resulting from tunneling are close to the phase modes described above. Then, J_{sp} can be calculated analytically. It is decreased from its value

$J_{\mathrm{sp}}^{(0)}$ at $N_{\mathrm{ex}} = 0$ and at the same hole concentration N_{h} by the following expression:

$$\frac{J_{\mathrm{sp}}}{J_{\mathrm{sp}}^{(0)}} = \frac{1}{2} \frac{|v_{k=0}^2|}{8\pi N_{\mathrm{ex}}} \leq \frac{1}{2}. \tag{4.41}$$

For smaller densities N_{ex}, or larger potentials, the approximation of phase modes cannot be used. In principle, spontaneous tunneling could still be analyzed within the pairing approximation, after the collective excitations have been calculated as in [136]. However, in the case $V - V_{\mathrm{min}} \ll 1$ (tunneling into excitons only) and for small N_{ex}, the interacting-Bose-gas-approximation becomes valid and is much more convenient to use. In this case, the two-body repulsive interaction between the excitons is treated within the T-matrix approximation to produce the new collective modes of the system [133]. The full derivation of the spontaneous tunneling rate can be found in [138], and here we report only the final result:

$$\frac{J_{\mathrm{sp}}}{J_{\mathrm{sp}}^{(0)}} = \frac{1}{2} \left(1 + \frac{\sqrt{(V - V_{\mathrm{min}})^2 + B^2} - B}{V - V_{\mathrm{min}}} \right), \tag{4.42}$$

where $B \sim -N_{\mathrm{ex}}/\log N_{\mathrm{ex}}$ is the low-momentum value of the anomalous exciton self-energy. At low bias voltages $V - V_{\mathrm{min}} \ll 1$, it is convenient to interpolate between the low-density results of (4.42) and the higher-density results of (4.41) with the following expression:

$$\frac{J_{\mathrm{sp}}}{J_{\mathrm{sp}}^{(0)}} = \frac{1}{2} \left(1 + \frac{\sqrt{(V - V_{\mathrm{min}})^2 + B^2} - B}{V - V_{\mathrm{min}}} \right) \frac{|v_{k=0}^2|}{8\pi N_{\mathrm{ex}}}. \tag{4.43}$$

We notice that this expression, as well as (4.42), correctly describes the low-density limit $N_{\mathrm{ex}} \to 0$, $J_{\mathrm{sp}}(E) \to J_{\mathrm{sp}}^{(0)}(E)$.

In Fig. 4.11 we plot the spontaneus (short-dashed line) and the total (long-dashed line) tunneling currents as a function of the exciton concentration at $V - V_{\mathrm{min}} = 0.03$, under the strengthened conditions (4.37) introduced above. The initial decrease of the spontaneous tunneling current as a function of N_{ex} by a factor of 2 is predicted by (4.42) within the picture of the interacting Bose gas, and is due to effects of exciton–exciton interactions, while its further reduction is described by (4.41). A final remark on the interacting-Bose-gas approximation is due: no saturation of the stimulated tunneling current is predicted, and the saturation current J_0 is exceeded at a density $N_{\mathrm{ex}} = 0.04$. Therefore, the approximation certainly becomes incorrect beyond this density. We finally notice that the pairing approximation also correctly describes the low-density limit of the stimulated tunneling rate as found in the interacting-exciton approximation. However, we remark again that it is much more complicated to find an approximate *spontaneous* tunneling rate within the pairing approximation at low densities, as explained above.

4.4 Resonant Tunneling
into Microcavity Exciton Polariton States

From the previous sections, we can conclude that the two-particle nature of the hole-assisted resonant tunneling process of electrons results in the selective creation of excitons with small transverse momentum at low bias. The anisotropic radiation pattern and the superradiative decay rate of those excitons can be used for high-efficiency, fast LEDs. Unfortunately, for GaAs QWs, the efficiency and speed of such LEDs are limited by the following two factors [139].

First of all, the tunneling process creates both orthoexcitons (optically active) and paraexcitons (optically inactive). Paraexcitons are first converted to orthoexcitons by means of exciton–exciton scattering or exciton–free-carrier scattering. Since the energy separation between paraexcitons and orthoexcitons is much less than 1 meV in GaAs QWs, they are created simultaneously by the tunneling process. In fact, half of the HH excitons created by the tunneling process are paraexcitons. Even if at low bias most of the HH excitons are created with small $k_\parallel$, only half of them (i.e. orthoexcitons) can superradiatively decay into the normal direction before momentum relaxation occurs. The other half of them (i.e. paraexcitons) cannot radiatively recombine until they are converted to orthoexcitons. Since the spin relaxation process is rather slow, by the time the paraexcitons have been converted to orthoexcitons, their $k_\parallel$ distribution has already evolved from the initial narrow distribution near $k_\parallel = 0$ to a broad thermal distribution. Hence the converted orthoexcitons have an isotropic radiation pattern.

Secondly, the large exciton mass in a GaAs QW limits the selectivity of the tunneling process for generating cold excitons. This is because the exciton dispersion curve $E_{ex}(k_\parallel) = \hbar^2 k_\parallel^2 / 2m_{ex}$ is quite flat near $k_\parallel \simeq 0$, owing to the large exciton mass m_{ex}, as shown in Fig. 4.12. Since the exciton energy is not sensitive to its transverse momentum $\hbar k_\parallel$, even at 4 K, the thermal k vector $k_T = \sqrt{2m_{ex}k_BT}/\hbar \simeq 4.2 \times 10^7$ m^{-1} is still larger than the photon momentum $k_c = 2\pi n_{eff}/\lambda \simeq 2.7 \times 10^7$ m^{-1}. Therefore the thermal energy and the inhomogeneous broadening of the QW exciton energy level limit the selectivity of this tunneling process for creating excitons with $k_\parallel \simeq 0$.

To overcome the above two problems, we have proposed to embed a resonant-tunneling diode in a high-Q planar microcavity. When the cavity photon frequency ω_c is tuned to the QW exciton emission frequency ω_{ex}, the orthoexciton state is strongly coupled to the cavity photon state and forms two exciton polariton states with energy $\omega_{ex} \pm \Omega$, where Ω is the exciton–photon coupling constant. However, the paraexcitons are not coupled to the cavity photon mode at all, and thus their energy remains $\hbar\omega_{ex}$. Hence the strong coupling of the orthoexciton state to the cavity photon state pushes its energy away from the energy of the paraexciton state. The energy separation between the paraexcitons and the orthoexciton polaritons near $k_\parallel = 0$ is equal to the exciton–photon coupling constant Ω, which can be up to 10

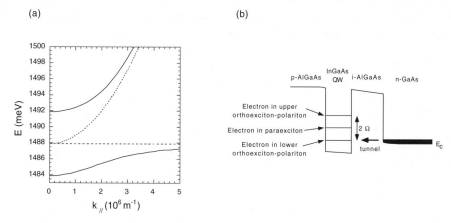

(a) (b)

Fig. 4.12. (a) *Solid lines:* calculated exciton polariton dispersion curves in a high-Q planar DBR microcavity. *Dashed line:* uncoupled exciton dispersion curve. *Dotted line:* uncoupled cavity photon dispersion curve. (b) Schematic diagram of the band structure of a resonant-tunneling diode embedded in a high-Q planar microcavity

meV in a high-Q microcavity [66]. Thus we can adjust the bias voltage so that electrons tunnel only into the lower polariton state, not into the paraexciton states (see Fig. 4.12b). In this way we can eliminate the creation of paraexcitons and increase the efficiency and speed of LEDs.

Moreover, since the microcavity exciton polariton state is a superposition of a QW exciton state and a cavity photon state, its mass is much smaller than the exciton mass. Therefore the exciton polariton dispersion curve $E_{\mathrm{pl}}(k_\parallel)$ is much steeper near $k_\parallel \simeq 0$, as shown in Fig. 4.12a. In other words, the energy of the exciton polariton state is much more sensitive to its $k_\parallel$ than that of the uncoupled exciton state. Thus, even at 77 K, the thermal k vector $k_T = \sqrt{2m_{\mathrm{pl}}k_{\mathrm{B}}T}/\hbar \simeq 1.8 \times 10^7$ m^{-1} is still smaller than the photon k vector $k_{\mathrm{c}} \simeq 2.7 \times 10^7$ m^{-1}. This leads to a significant improvement of the selectivity of this tunneling process for generating $k_\parallel \simeq 0$ exciton polaritons.

The matrix element for tunneling into a polariton state is equal to the tunneling matrix element into the exciton state multiplied by the exciton probability in the polariton state. This probability is one-half at resonance. The percentage of lower polaritons (LPs) created through resonant tunneling at a bias voltage V and with $k_\parallel \leq \sqrt{2m_{pl}\Omega}/\hbar$ is estimated to be [139]

$$\xi(V) \sim \left\{ 1 + 8\frac{m_{\mathrm{ex}}}{m_{\mathrm{pl}}} \left[\frac{k_{\mathrm{B}}T}{e(V - V_{\mathrm{min}})} \exp\left(\frac{e(V - V_{\mathrm{min}}) - \Omega}{k_{\mathrm{B}}T} \right) \right.\right.$$
$$\left.\left. + \frac{1}{2\pi} \left(\frac{E_{\mathrm{fe}} + |E_{\mathrm{fh}}|}{e(V - V_{\mathrm{min}})} \right)^2 \frac{\gamma}{\Omega} \right] \right\}^{-1}, \tag{4.44}$$

where E_{fe} and E_{fh} are the electron and hole Fermi energies, and V_{min} is the minimum bias voltage required to tunnel into the LP state.

To further increase the selectivity of this resonant tunneling process, we have proposed to insert this tunneling diode into a micropost cavity whose lateral size is less than 1 µm. Such a micropost structure (also called a pillar microcavity) provides a strong lateral confinement of the optical field. Owing to the large refractive-index difference between the semiconductor post and the surrounding air, the transverse momentum of the cavity photon state is quantized. This leads to discrete transverse photon modes [140, 141]. As the lateral dimensions shrink, the transverse momenta of these discrete cavity photon modes are moved further apart.

In the strong-coupling regime, in a planar microcavity, the transverse momentum of the exciton polaritons can have any value. Therefore there is a continuum of exciton polariton states in the planar microcavity. However, in a pillar microcavity, the transverse momentum of the exciton polariton states is quantized. The $k_{\parallel}$ separation between these discrete exciton polariton states increases as the lateral size of the microcavity decreases. For example, if the lateral size of the pillar microcavity is about 0.5 µm, the $k_{\parallel}$ separation between the first and second quantized LP states leads to a large energy separation between them, owing to the strong LP dispersion near $k_{\parallel} = 0$. This enables us to adjust the bias voltage such that the electrons tunnel only into the first LP state (the one with the lowest energy). The emission efficiency of this light-emitting device is calculated to be more than 50% [139].

Therefore, the hole-assisted resonant tunneling process of electrons provides a direct, efficient electrical pumping scheme for microcavity exciton polaritons.

5. Competition Between Photon Lasing and Exciton Lasing

5.1 Bose–Einstein Condensation of Excitons

The impressive progress in laser cooling and trapping of atoms has recently resulted in the experimental realization of Bose–Einstein condensation (BEC) of cold rubidium, lithium, and sodium atoms [142–144]. Since excitons are approximately bosons, BEC of excitons or excitonic molecules has attracted a lot of interest [145]. In bulk crystals, degenerate Bose–Einstein (BE) statistics for excitons in Ge [146] and biexcitons in CuCl [147] have been reported. BE saturation in Cu_2O has also been observed [148].

One of the important factors that makes it difficult to show direct evidence of BEC in any exciton system is the fact that the excitonic lifetime is usually short. For example, the lifetime of excitons in direct-gap materials is usually not sufficiently long compared with the thermalization time, and thus it prevents the formation of a condensed state. The exciton lifetime in indirect-gap materials is longer than the thermalization time, but excitons condense into electron–hole plasma droplets before BEC. Cuprous oxide is a promising candidate for BEC of excitons owing to its forbidden direct gap, the large excitonic binding energy (150 meV) [149], and the effect of electron–hole exchange, which inhibits the formation of complexes such as biexcitons or electron–hole droplets. In 1993, Lin and Wolfe reported an experimental observation of BEC of paraexcitons in stressed Cu_2O [150].

In a strictly two-dimensional system, although ideal BEC cannot occur, a Kosterlitz–Thouless transition can be expected. For a two-dimensional system of spatially separated electron–hole pairs, Lozovik and Yudson [151] and Shevchenko [152] independently suggested the possibility of a phase transition into an ordered state. Their proposal, based on an analogy with BCS theory, has been extended to a coupled quantum well system subject to an electric field, which allows the realization of a long-lived two-dimensional exciton system [153]. Unfortunately, in such systems, disorder dominates the luminescence properties at low densities. As a consequence, condensation over any macroscopic region of space is unlikely, although evidence for "puddles" of indirect exciton condensate has been inferred in coupled AlAs/GaAs quantum wells [154].

5.2 Exciton Lasers (Bosers)

There have been several proposals for atom [155,156] and exciton [157] lasers. In contrast to Bose–Einstein condensates [158–161], atomic or excitonic systems are driven far from equilibrium in these matter wave lasers [162].

What is exciton boser? As we know, for a massive bosonic field, quantum statistics become important when the thermal de Broglie wavelength exceeds the interparticle spacing. An exciton boser is a device that utilizes quantum statistical effects to generate a coherent population of nonequilibrium excitons. The fundamental physical process that results in coherent-state formation in a boser is the many-body exciton–phonon interaction. Owing to the bosonic nature of excitons, the spontaneous phonon emission rate into the lowest-energy excitonic state is enhanced by *final-state stimulation*, which in turn results in an exponential growth of occupancy.

Let us consider an exciton boser, where we assume that all but the $k = 0$ excitons form an excitonic reservoir. This (electrically neutral) reservoir is pumped by an external incoherent source (either electrical injection or light), and therefore is out of equilibrium with the phonon reservoir that it interacts with. We assume that the excitation intensity is weak enough that the excitons (both those in the reservoir and those with $k = 0$) may be treated as bosonic particles. A simple rate equation can be set up to estimate the exciton boser threshold:

$$\frac{d}{dt}N_0 = -\frac{N_0}{\tau_0} + \sum_k \Gamma_{\mathrm{ph}}(k)[N_k(N_0 + 1)(n_k + 1) - N_0(N_k + 1)n_k],$$

$$\frac{d}{dt}N_k = -\frac{N_k}{\tau_k} - \sum_k \Gamma_{\mathrm{ph}}(k)[N_k(N_0 + 1)(n_k + 1) - N_0(N_k + 1)n_k].$$

$$(5.1)$$

N_0 represents the exciton occupancy in the $k = 0$ state, N_k represents the exciton occupancy in the momentum state $\hbar k$ in the exciton reservoir, and n_k represents the phonon occupancy in the momentum state $\hbar k$ in the phonon reservoir. Γ_{ph} is the exciton–phonon scattering rate. The first term on the right-hand side of (5.1) represents the radiative decay of excitons, and the second term corresponds to the transition of an exciton from a $k \neq 0$ state to the $k = 0$ state by emitting a phonon. The transition rate is proportional to N_k and $N_0 + 1$. The third term represents the opposite process, i.e. the transition of an exciton from the $k = 0$ state to a $k \neq 0$ state by absorbing a phonon. Hence the transition rate is proportional to N_0 and $N_k + 1$. Therefore the second term represents the stimulated generation of $k = 0$ excitons through phonon emission processes, and the third term represents the absorption of $k = 0$ excitons through phonon absorption. To obtain a net gain for the $k = 0$ excitons, the second term must be larger than the third term. This requires that

$$N_k > n_k.$$

$$(5.2)$$

This condition indicates that for all modes that contribute to $k = 0$ exciton–phonon scattering, the exciton occupancy has to exceed the phonon occupancy. This is the *boser inversion* condition. The actual observation of population buildup in the $k = 0$ excitonic state also requires that the (unsaturated) *net gain* exceeds the loss, i.e. the boser threshold is

$$\sum_k \Gamma_{\text{ph}}(k)\,(N_k - n_k) \geq \Gamma_{\text{loss}} \,, \tag{5.3}$$

where the sum is over all excitonic modes, and $\Gamma_{\text{loss}} = 1/\tau_0$. If (5.3) is satisfied, a coherent state of excitons in the $k = 0$ state with a slowly diffusing phase will form. A numerical simulation based on a stochastic-wavefunction approach has shown a transition from a thermal state to a coherent state in an exciton boser [157]. This indicates a mean exciton field will be generated in an exciton boser. The phase fluctuations have contributions from spontaneous exciton–phonon scattering and exciton–exciton interactions, both of which remain significant at $T_{\text{ph}} = 0$ K. The linewidth reduction obtained in a boser is limited by the strength of these interactions. We have assumed here that the phonon reservoir remains in a thermal state with a well-defined temperature T_{ph}. In contrast to BEC, the dimensionality of the excitonic system is not important in an exciton boser.

The boser inversion condition (5.2) corresponds to the minimum reservoir exciton density $N_{\text{exc}}^{\text{min}}$ required for there to be net final-state stimulated emission of ground-state excitons

$$N_{\text{exc}}^{\text{min}} \simeq 2.62\,\lambda_{T_{\text{ph}}}^{-3} \,, \tag{5.4}$$

where $\lambda_{T_{\text{ph}}} = \sqrt{2\pi\hbar^2/(m_{\text{ex}}kT_{\text{ph}})}$ denotes the thermal de Broglie wavelength of an exciton gas at temperature T_{ph} ($= T_{\text{exc}}$ in thermal equilibrium). This indicates that the observation of final-state stimulation requires an exciton density which exceeds the minimum density required for BEC.

Extension of (5.4) to a two-dimensional exciton gas poses no difficulties: since phonons with $\omega < \omega_{\text{min}} = 2m_{\text{ex}}c_{\text{s}}^2/h$ do not contribute to phonon absorption/emission, the divergence of the phonon number as $\omega \to 0$ is irrelevant in the sense that $N(\omega) > n(\omega)$ needs only to be satisfied for $\omega > \omega_{\text{min}}$. Therefore, even a thermal distribution of reservoir excitons ($N(\omega) = \{\exp[(\hbar\omega - \mu)/(kT_{\text{exc}})] - 1\}^{-1}$) with chemical potential $\mu < 0$ and temperature $T_{\text{exc}} \geq T_{\text{ph}}(1 - \mu/\hbar\omega_{\text{min}})$ will be sufficient to obtain net stimulated emission of excitons.

The excess reservoir exciton density that is required for the formation of a nonequilibrium condensate or an exciton boser depends on the loss rate Γ_{loss} and the frequency dependence of the phonon emission/absorption rate $\Gamma_{\text{ph}}(\omega)$. If we assume a hypothetical system where $\Gamma_{\text{ph}}(\omega) \simeq \Gamma_{\text{ph}}$, we obtain

$$N_{\text{exc}}^{\text{thres}} = \frac{2.62}{\lambda_{T_{\text{ph}}}^3} + \frac{\Gamma_{\text{loss}}}{V\,\Gamma_{\text{ph}}} \,, \tag{5.5}$$

as the density required to obtain a nonequilibrium exciton condensate. Here $V \geq \lambda_{T_{\mathrm{ph}}}^3$ denotes the volume of the semiconductor (quantization volume) that is assumed to be free of defects and impurities. Physically, $N_{\mathrm{exc}}^{\mathrm{thres}}$ corresponds to the reservoir density at which the net gain equals the net loss, per unit volume. In the equilibrium limit ($\Gamma_{\mathrm{loss}} \to 0$), we naturally obtain the requirement for BEC.

We would like to comment that the density requirement of (5.5) (obtained from a frequency-independent phonon scattering rate) is sufficient but not necessary in many particular realizations. If we consider, for example, the case where the principal gain mechanism is final-state stimulation through longitudinal optical (LO) phonon emission, the only requirement is a large exciton reservoir occupancy at $\omega_{\mathrm{exc}} = \omega_{\mathrm{ph}}$, since the LO-phonon scattering rate is much larger than the acoustic-phonon scattering rate, i.e. $\Gamma_{\mathrm{LO}} \gg \Gamma_{\mathrm{ph}}$. In fact, the exciton occupancy at other frequencies may be well below that of phonons, so that $N_{\mathrm{exc}} \ll 2.62 \lambda_{T_{\mathrm{ph}}}^{-3}$. Therefore, provided that we can keep the exciton reservoir far from equilibrium, condensation effects may be observed at densities lower than that required for BEC.

It has been shown theoretically that when the above conditions are satisfied, a coherent state of excitons with a diffusing phase is formed [157]. The output from such a nonequilibrium condensate is either a coherent matter wave (obtained when the condensed excitons are allowed to tunnel out into another semiconductor medium) or a coherent light wave (obtained when the coherent excitons hit a physical boundary, or when there is a finite final-radiation-field density of states). Since conventional coherent light sources depend on the existence of an electronic population inversion, it is important to determine if a similar condition is necessary for an exciton boser.

The semiconductor Bloch equations specify the relevant inversion operator for generation of coherent light from an incoherent (uncorrelated) electron–hole reservoir as

$$\hat{I}_{\boldsymbol{p}} = 1 - \hat{n}_{\mathrm{e},\boldsymbol{p}} - \hat{n}_{\mathrm{h},-\boldsymbol{p}} \,, \tag{5.6}$$

where $\hat{n}_{\mathrm{e},\boldsymbol{p}}$ and $\hat{n}_{\mathrm{h},-\boldsymbol{p}}$ denote the electron and hole number operators, respectively. If $\langle \hat{I}_{\boldsymbol{p}} \rangle > 0$ for all $\boldsymbol{p}$, all applied (weak) electromagnetic fields will experience net loss [163].

Light generation at the ground-state exciton frequency is achieved by a superposition excitation of free electron–hole pairs, i.e. the radiation field reservoir mode $\hat{a}_{\boldsymbol{k}}$ couples to the exciton mode $\hat{b}_{\boldsymbol{k}}^\dagger$ with

$$\hat{b}_{\boldsymbol{k}}^\dagger = \frac{1}{\sqrt{V}} \sum_{\boldsymbol{p}} \varphi(p) \hat{e}_{\boldsymbol{k}/2+\boldsymbol{p}}^\dagger \hat{h}_{\boldsymbol{k}/2-\boldsymbol{p}}^\dagger \,, \tag{5.7}$$

where $\hat{e}_{\boldsymbol{p}}^\dagger$ and $\hat{h}_{-\boldsymbol{p}}^\dagger$ denote the electron and hole creation operators, and

$$\varphi(p) = 8\sqrt{\pi a_{\mathrm{B}}^3} \frac{1}{[1 + (p a_{\mathrm{B}})^3]^2} \,, \tag{5.8}$$

is the Fourier transform of the hydrogenic 1s wavefunction. The average inversion operator is $\hat{\bar{I}} = (1/V) \sum_{\boldsymbol{p}} |\varphi(p)|^2 \hat{I}_{\boldsymbol{p}}$, and its expectation value is

$$\langle \hat{\bar{I}} \rangle \geq 1 - 2a_{\mathrm{B}}^3 \left(N_{\mathrm{exc}} + \frac{n_0}{V} \right) , \tag{5.9}$$

where n_0 is the average ground-state exciton occupancy. Provided that $N_{\mathrm{exc}} + n_0/V)a_{\mathrm{B}}^3 \ll 1$, there is no electronic inversion in the system, i.e. $\langle \hat{\bar{I}} \rangle > 0$. This result is not unexpected, since $\hat{\bar{I}} = [\hat{b}, \hat{b}^\dagger]$, and the bosonic character of the excitons implies the absence of electronic inversion.

Coherent light generation from an exciton boser requires in most cases $N_{\mathrm{exc}} \lambda_{T_{\mathrm{ph}}}^3 > 2.62$. At a low enough temperature, where $\lambda_{T_{\mathrm{ph}}} \gg a_{\mathrm{B}}$, $N_{\mathrm{exc}} a_{\mathrm{B}}^3 \ll 1$ are satisfied, all optically active electron–hole sates are noninverted.

Since a coherent exciton state generates coherent light by spontaneous radiative recombination, an exciton boser may be viewed as a *laser without inversion* when inversion is defined as $\langle \hat{\bar{I}} \rangle < 0$. There is a significant difference between the exciton boser described here and the II–VI exciton laser, where a BCS-type state of excitons is utilized for coherent light generation [164]; a BCS state of excitons is electronically inverted.

5.3 Exciton Polariton Bosers

Unfortunately, in GaAs, the exciton Bohr radius is quite large, i.e. $a_{\mathrm{B}} \sim 10$ nm. Even at 4 K, the exciton thermal de Broglie wavelength λ_T is still comparable to the exciton Bohr radius. In order to increase λ_T, we can dress the exciton with a photon in a semiconductor microcavity to form a new quasiparticle, the exciton polariton:

$$\hat{p}_{\boldsymbol{k}} = u_{\boldsymbol{k}} \hat{b}_{\boldsymbol{k}} + v_{\boldsymbol{k}} \hat{a}_{\boldsymbol{k}} , \tag{5.10}$$

where $\hat{p}_{\boldsymbol{k}}$, $\hat{b}_{\boldsymbol{k}}$ and $\hat{a}_{\boldsymbol{k}}$ denote the polariton, exciton, and photon annihilation operators with wavevector $\boldsymbol{k}$, respectively. Since the microcavity exciton polariton state is a superposition of a QW exciton state and a cavity photon state, the exciton polariton mass is much smaller than the exciton mass, and thus its thermal de Broglie wavelength is much larger. For instance, at 4 K the thermal de Broglie wavelength of a microcavity exciton polariton is about 7 μm, as compared with 0.07 μm for a bare exciton. Hence, the threshold of an exciton polariton boser is much lower than that of an exciton boser. Moreover, owing to their light mass (small density of final states), microcavity exciton polaritons have also a smaller rate of scattering by acoustic-phonon absorption [165]. GaAs QW excitons are subject to exciton–exciton interaction and phase-space-filling effects when the interparticle spacing is less than 0.3 μm and 0.03 μm, respectively [100]. These critical values for undesirable scattering are also improved by dressing the excitons with a vacuum electromagnetic field. In short, it is much easier to achieve an exciton polariton boser than an exciton boser.

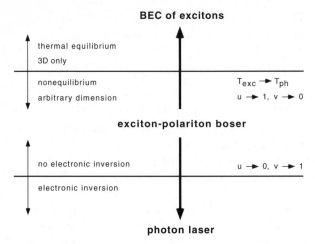

Fig. 5.1. Diagram comparing the exciton polariton boser with a Bose–Einstein condensate of excitons and a photon laser

An exciton polariton boser simultaneously generates a coherent exciton (matter) wave and a coherent optical wave. Figure 5.1 compares the exciton polariton boser with Bose–Einstein condensate of excitons and a photon laser. For a reservoir in thermal equilibrium and a vanishing photon character of the exciton polaritons ($v \rightarrow 0$), one obtains a BEC of excitons, provided that the system is three-dimensional. In the opposite limit of a nonequilibrium (inverted) reservoir and a vanishing exciton character ($u \rightarrow 0$), the exciton polariton boser is indistinguishable from a photon laser.

In a below-threshold boser, where $0 < \Gamma_{\text{gain}} < \Gamma_{\text{loss}}$, an applied weak probe field will experience net gain despite the fact that there is no electronic population inversion. In this limit, the exciton polariton boser can be regarded as a regenerative (photon) laser amplifier without inversion, which utilizes a many-body coherence within the electron–hole system. This is in contrast to other inversionless laser schemes, which depend on single-particle coherence [166–169].

5.4 Pump–Probe Experiments

Experimentally, we have tried to realize an exciton polariton boser in a high-Q semiconductor microcavity.

The sample used in these experiments was grown by molecular-beam epitaxy and consisted of a single 20 nm GaAs quantum well in a half-wavelength DBR cavity. The cavity buffer layer was tapered along one direction so that the cavity resonant frequency varied with sample position. The sample was cooled to 4.2 K in a liquid-helium cryostat. To avoid time-dependent effects, a

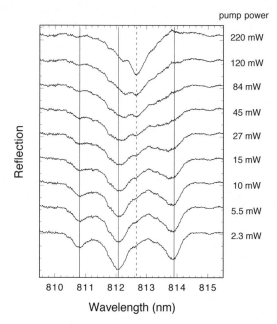

pump power

220 mW

120 mW

84 mW

45 mW

27 mW

15 mW

10 mW

5.5 mW

2.3 mW

Reflection

810   811   812   813   814   815

Wavelength (nm)

Fig. 5.2. Reflection spectra of the probe at different incident pump powers

continous-wave (CW) Ti:sapphire laser, operating at 767 nm, was used as the pump. It was focused to a 30 μm spot on the sample. The cavity emission into the normal direction was measured by a spectrometer. A weak probe beam from a mode-locked Ti:sapphire laser, operating at 810 nm, was incident onto the central part of the pump spot on the sample. The reflected probe beam was guided to a spectrometer.

We measured the cavity emission and the probe reflection as we increased the pump power. The sample position was chosen to be where the cavity photon energy was close to the QW HH exciton energy. Figure 5.2 shows the reflection spectra of the probe at different pump powers. The QW exciton density corresponding to 1 mW pump power is about 2×10^9 cm^{-2}. At low pump power, the strong coupling of both QW HH excitons and QW LH excitons to the cavity photon state results in three exciton polariton states, corresponding to the three dips in the reflection spectrum. As the pump power increases, the exciton polariton peaks become broadened, and the normal-mode splitting also reduces slightly. Eventually, at high enough power, the system makes a continuous transition to the weak-coupling regime, and the reflection spectrum has only one dip, which corresponds to the bare cavity photon mode. Figure 5.3 shows the emission spectra taken simultaneously at the same pump powers. At low pump power, the emission spectrum features two HH exciton polariton peaks. The slight blue shift of the exciton polariton frequency in the reflection spectrum as compared with the emission spectrum

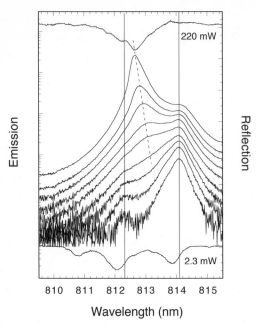

Fig. 5.3. Cavity emission spectra at the same pump powers as those in Fig. 5.2. Two reflection spectra at low and high pump powers also shown for comparison

is due to the 5° incident angle of the probe beam. As the pump power increases, a third peak emerges in between the two exciton polariton peaks, and its intensity grows nonlinearly. By comparing with the reflection spectrum, it can be seen that the third emission peak appears at a wavelength close to that of the cavity photon mode. This indicates that the system starts lasing in the cavity photon mode. Owing to the nonuniform spatial distribution of the light intensity over the pump spot, the probe reflection spectrum changes from a single cavity photon mode to three exciton polariton modes as we move the probe spot from the pump spot center to its edge. This indicates that the excitons are saturated in the central part of the pump spot, but not in the edge. In the emission spectrum, the lasing line is from the center of the pump spot, and the lower HH exciton polariton peak is from the edge of the pump spot. This was confirmed experimentally by the fact that the lower HH exciton polariton emission peak was suppressed significantly when we measured the emission from only the central part of the pump spot.

5.5 Effective-Mass Measurement

We also measured the cavity emission spectra at different directions and pump powers. Figure 5.4 shows the energies of the two HH exciton polaritons

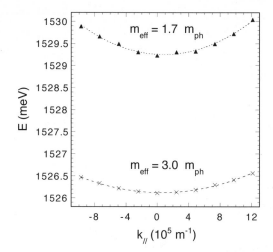

Fig. 5.4. Energies of the two HH exciton polaritons as a function of $k_\parallel$ at an incident pump power of 2.3 mW

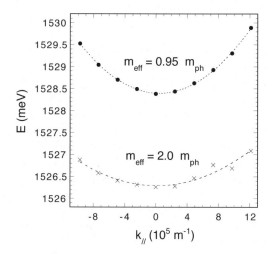

Fig. 5.5. Energies of the lasing line (*filled circles*) and of the lower HH exciton polariton (*crosses*) as a function of $k_\parallel$ at an incident pump power of 120 mW

as a function of $k_\parallel$ at low pump power. From the second derivative of their angular dispersion curves at $k_\parallel = 0$, we deduced the effective masses of the HH exciton polaritons to be $1.7m_{\mathrm{ph}}$ and $3.0m_{\mathrm{ph}}$, where m_{ph} is the bare cavity photon mass. Note that exciton polaritons at an exact anticrossing point should have an effective mass of $2m_{\mathrm{ph}}$. This indicates that the cavity photon energy is slightly blue-shifted with respect to the HH exciton energy. Figure 5.5 shows the angular dispersion curves of the lasing line and of the lower HH exciton polariton at high pump power. The effective mass of the lasing line

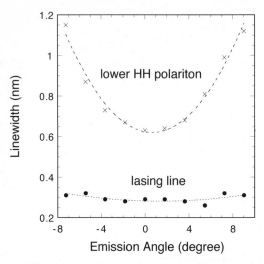

Fig. 5.6. Linewidth of the lasing line (*filled circles*) and of the lower HH exciton polariton peak (*crosses*) as a function of the emission angle (in the air) at an incident pump power of 120 mW

is very close to the bare cavity photon mass, which confirms the bare-photon nature of the lasing line. The slight decrease in the lower-polariton effective mass may be due to the blue shift of the HH exciton energy.

Figure 5.6 shows the linewidth of the lasing line and of the lower HH exciton polariton peak as a function of emission angle at high pump power. The linewidth of the lower HH exciton polariton peak increases as $k_\parallel$ increases, because excitons with larger $k_\parallel$ not only radiatively decay but also relax to smaller-$k_\parallel$ states by acoustic-phonon emission. However, the linewidth of the lasing line is almost independent of the emission angle; this is a characteristic of a photon laser. The intensity of the lasing line drops rapidly as the emission angle increases, as shown in Fig. 5.7. The half angle of the emission lobe is about 8° in the air, which is close to the divergence angle of the bare cavity photon mode as determined by the Q value of the cavity [171, 172]. Therefore the microcavity system indeed behaves like a photon laser at high pump power.

We have repeated this experiment at many other sample positions and observed similar lasing phenomena. Although at different sample positions the cavity exciton detuning is different, and hence the effective masses of the exciton polaritons vary a lot, the lasing line features a universal effective mass, which is the bare cavity photon mass. As an example, Figs. 5.8 and 5.9 show data taken at another sample position. From the evolution of the cavity emission spectra as a function of the pump power, it seems that no additional emission peak emerges, and the intensity of the middle exciton polariton grows nonlinearly. This looks like a spontaneous buildup of the polariton population, which might originate from the stimulated generation of

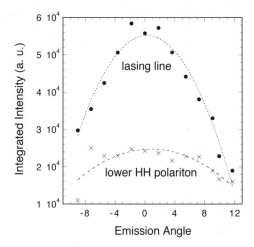

Fig. 5.7. Integrated intensity of the lasing line (*filled circles*) and of the lower HH exciton polariton peak (*crosses*) as a function of the emission angle (in the air) at an incident pump power of 120 mW

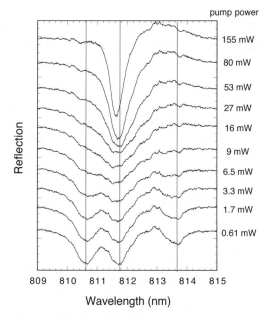

Fig. 5.8. Reflection spectra of the probe at different incident pump powers, at a different sample position from that used in Fig. 5.2

coherent exciton polaritons through phonon-scattering processes [170]. However, the simultaneous measurement of the reflection spectra shows that, as the pump power increases, the QW excitons become saturated and the exciton polaritons eventually disappear. At high pump power, the reflection

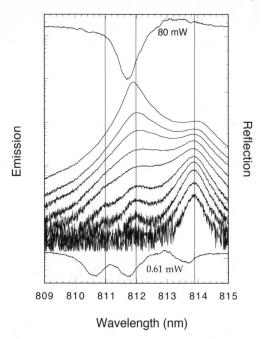

Fig. 5.9. Cavity emission spectra at the pump powers corresponding to those in Fig. 5.8, at the same sample position. Two reflection spectra at low and high pump power also shown for comparison

spectrum shows only the bare cavity photon mode, which is very close to the wavelength of the middle exciton polariton at low pump power.

In summary, we have observed a continuous transition passing directly from exciton polariton emission to a bare-photon laser without going through the intermediate phase of an exciton polariton boser. This suggests several defects in the present experiment. In our sample, the size of the polariton center-of-mass wavefunction is much smaller than the polariton thermal de Broglie wavelength (calculated using the measured quasiparticle mass), owing to exciton localization caused by the QW interface roughness and impurities. The reduction of the actual size of the polariton wavefunction along the QW plane increases the exciton polariton laser threshold density to above the exciton saturation density. The nonresonant pumping creates a lot of free carriers and hot excitons, which scatter the ground state ($k_\parallel = 0$) polariton effectively and destroy its coherence. Finally, the polariton lifetime is too short (~ 1 ps) compared with the relatively slow exciton thermalization due to phonon-scattering processes (~ 1 ns). This slow thermalization of excitons due to the so-called "bottleneck effect" will be discussed in the next chapter, where a new scheme to realize stimulated emission of polaritons based on exciton–exciton scattering is proposed.

6. Polariton Dynamics in Photoluminescence

6.1 Introduction

In this chapter we discuss the dynamics of polaritons in photoluminescence experiments. In contrast to the case of resonant, pulse-like excitation described in Sect. 2.2, here we consider nonresonant optical excitation of the system. The energy of the photons used for such excitation is usually well above all the energies of the observed excitations and, in particular, of the exciton polaritons. Usually, free electron–hole pairs are created by the excitation process, as the exciton binding energy amounts to some tens of meV only. It is usually assumed that these pairs cool down, and form hot excitons (bound electron–hole pairs), by emitting phonons. This part of the dynamics is not trivial, but is not relevant to the subsequent exciton polariton dynamics itself. Moreover, the issue of exciton formation is still controversial. Excitons are certainly formed at long times after excitation, but some exciton-like emission from the free carriers apparently starts at much shorter times owing to the Coulomb correlations between the charged carriers in the emission process [173, 174]. Establishing the weights of these two processes – i.e. the part of the excitonic luminescence originating from formed, bound excitons or that originating from excitonic recombination of free carriers – is well beyond our purpose. A simple rate equation analysis indicates a strong nonequilibrium situation biased towards free carriers at low temperatures in typical III–V materials [175], owing to long formation times of excitons compared with their short radiative lifetime. If one wishes to focus on the exciton polariton dynamics only, excitons should be formed directly in experiments by resonant excitation. This is possible, for example, by using lasers at large incidence angles on a microcavity, as we explain later. In this case, free carriers are unlikely to be formed at low temperatures, as this involves absorption of many acoustic phonons. In the following we restrict our analysis to this case only.

6.2 Kinetic Equations

The dynamic processes which we have to consider are exchange of energy with the lattice through scattering with acoustic phonons, binary interparticle scattering, and escape from the microcavity through emission of photons. Scattering from impurities and surface roughness is elastic, and does not modify the population distribution of the exciton polaritons. We shall discuss this point later in more detail. We treat the dynamics by a kinetic approach, in which the renormalization of quasiparticle energies due to the interactions is completely neglected, and only the scattering rates are considered. This description results in a kinetic Boltzmann equation for the exciton polariton bosonic fluid. When the Boltzmann equation is discretized, it results in a finite set of rate equations. A natural discretization of the modes of the system results when the system is enclosed in a large box. The quasiparticle momenta are then quantized into a uniform mesh. For our system, this has practical drawbacks. Exciton-like polaritons and strongly coupled polaritons have very different masses. A dense grid in k space for the light-mass polariton would result in a grid for the heavy-mass exciton-like polaritons having a prohibitively large number of k points. As an alternative to this "canonical" discretization, we may bin the energies of the quasiparticles into a uniform mesh. This has additional advantages. The uniform energy grid allows us to impose exactly the energy conservation condition for elastic exciton–exciton scattering, avoiding drifts in the energy conservation in the numerical integration of the rate equations. The energy grid is defined as

$$E_i^{(j)} = E^{(1)}(k = 0) + (i + 1/2)\Delta E, \qquad (6.1)$$

with $i = 0, 1, \ldots$, $j = 1, 2$. Here ΔE defines the energy spacing, $j = 1$ labels the lower branch, and $j = 2$ the upper branch. We have used the same energy grid for both upper and lower polaritons, which makes it immediately possible to write the condition for energy conservation in the elastic scattering. In the following, we shall use one index only, to label both the energy bin and the branch, for compactness. The energy grid defines a nonuniform k grid, $\{k_i\}$, such that $E(k_i) = E_i$.

The choice of ΔE is influenced by many factors. First, the population distributions have to be described accurately. Thus, the population should vary by only a small amount over ΔE. This gives the restriction $\Delta E < k_\mathrm{B}T$ for a classical thermal distribution, but becomes more stringent when bosons show quantum degeneracy. Second, we need a good description of the scattering-matrix elements with phonons. The typical energy exchanged by excitons with phonons is 1 meV in GaAs-based materials. This gives the restriction $\Delta E \ll 1$ meV. Third, we need a good description of the exciton–exciton scattering matrix elements. However, the elastic scattering becomes singular for small exchanges of energy. This problem is related to the validity of using a Boltzmann equation for describing the elastic scattering. Let us therefore review this point. In the kinetic approach, successive scatterings are assumed to be independent of each other. The quasiparticles are freely moving most of the time, and experience short binary collisions once in a while. The kinetic approach thus does not describe how phase information is carried over from one scattering to the next one by a freely moving particle. Only the population of the states is described, and scattering rates are calculated within the Fermi golden rule, which assumes the asymptotic states to be free. The approximation of independent scatterings holds – both for elastic and for inelastic scattering – when $\hbar^2 k \Delta k/m \gg \Gamma$, where k is the momentum of the excitation considered, Δk is its change in the collision, and Γ is the collision rate [176]. The semiclassical approach becomes inaccurate for small momenta, unless the scattering rate scales to zero faster than k. This is not the case for either phonon or exciton–exciton collisions. The region where the semiclassical approximation breaks down for phonon scattering is rather negligible, unless degenerate bosons are considered. But it can be of considerable importance in the case of exciton–exciton scattering at large densities. Moreover, in this case, the scattering rate also diverges for small exchanged momenta, even for finite k. For these particular collisions we may not apply the Boltzmann equation. In particular, strong renormalization at small momenta is responsible for the onset of superfluidity of interacting bosons [177]. However, when we discretize the energy, the collisions where small momenta are exchanged are also mostly within one energy bin, and are thus not apparent in the dynamics of the discretized equation. Thus, ΔE sets a minimum cutoff energy that we need to consider in the dynamics. The rate equations then hold when $\Delta E \gg \hbar\Gamma$, where Γ now is a typical exciton–exciton scattering rate. A meaningful estimate is the total scattering rate out of a given bin, which is mildly dependent on the shape of the population distribution, and given by

$$\hbar \Gamma_{\mathrm{exc-exc}} = \frac{\pi^2}{2} (n_{\mathrm{exc}} a_{\mathrm{B}}^2) \frac{E_{\mathrm{B}}^2}{\hbar^2/m_{\mathrm{exc}} a_{\mathrm{B}}^2} \tag{6.2}$$

for a classical thermal distribution. In a typical GaAs QW, $\hbar\Gamma \sim 1$ meV at $n_{\mathrm{exc}} = 10^{10}$ cm^{-2}, and the rate equations certainly hold up to $n_{\mathrm{exc}} \sim 10^9$ cm^{-2}. At larger densities, we may still retain these results, bearing in mind that they are only indicative, and missing the physics of the quasiparticle energy renormalization, and eventually the onset of coherent transport phenomena such as superfluidity.

The discretized rate equations read

$$\dot{N}_i = P_i - \Gamma_i N_i - W_{i,i'} N_i (N_{i'} + 1) + i \leftrightarrow i'$$
$$-Y_{ii_1,i'i'_1} N_i N_{i_1} (N_{i'} + 1)(N_{i'_1} + 1) + \left\{ \begin{matrix} i \leftrightarrow i' \\ i_1 \leftrightarrow i'_1 \end{matrix} \right\}, \tag{6.3}$$

where repeated indices are summed. Here Γ_i are the radiative recombination rates, $W_{i,i'}$ are the scattering rates of polaritons with phonons, $Y_{ii_1,i'i'_1}$ are the exciton–exciton scattering rates, N_i is the population at k_i, and P_i is the pumping rate. Next, we discuss the calculation of Γ_i, $W_{i,i'}$, and $Y_{ii_1,i'i'_1}$. In order to calculate the scattering rates, we need to know the dispersion of the exciton polaritons. This was introduced in Sect. 2.1, and can be accurately calculated within a two-oscillator model [178].

6.2.1 Radiative Recombination of Exciton Polaritons

Radiative recombination of exciton polaritons results from the leakage of the cavity photons through the cavity mirrors. This can be taken into account in the two-oscillator model by introducing an imaginary component into the cavity photon eigenenergies which equals half the escape rate of the photon from the cavity. This results in an imaginary component of the exciton polariton eigenenergy which is half the radiative recombination rate. This physical picture is, however, correct only at small angles, when the mirror reflectivity is large and close to one. At larger angles (or larger in-plane k vectors), the DBR reflectivity drops to small values, as shown in Sect. 2.1.1, and the mirrors become severely leaky. The leaked photons escape into the substrate, which has a large refractive index, similar to that of the cavity. A detailed calculation of the reflectivity as a function of the angle is therefore necessary in order to obtain meaningful radiative recombination rates. The resulting radiative recombination rate of the (mainly) exciton-like polaritons thus shows a rich structure [179]. The average of the radiative recombination rate in this leaky-mode region does not markedly differ from the average of the radiative recombination rate of a bare QW exciton. Thus, it is unlikely that the fine details show up in the photoluminescence dynamics, where the exciton population distribution spans a large phase space, several times larger than the radiative region $k < \omega/v$. However, we want to accurately determine the recombination rates in the transition region between large angles and small angles, where exciton-like polaritons become increasingly admixed with cavity photons. The population found in this transition region depends, in fact, on these radiative rates and, as we shall later show, can potentially become larger than one, thus featuring stimulated emission. For this reason, we carried out a detailed calculation of the recombination rate including all the structure of the DBR mirrors, as explained in [179]. We then averaged the resulting rates over the discrete energy grid defined in (6.1), using a very small value of $\Delta E = 0.05$ meV. The results for a $Ga_{0.7}Al_{0.3}As\lambda$ microcavity embedding an 80 Å QW at its center, enclosed by a DBR of 16 $\lambda/4$ pairs at the top (facing the air), and a DBR of 24 pairs at the bottom (facing the substrate), are shown in Fig. 6.1. There, we also show the corresponding dispersion and the quantized energy grid.

6.2.2 Polariton–Phonon Scattering

The interaction of electrons and holes with acoustic phonons in GaAs mainly originates from the rigid shift of the bands which occurs when hydrostatic pressure is applied to the crystal. This shift is characterized by a deformation potential, which is the shift of the band under consideration for unity strain. The exciton–phonon interaction matrix elements are then simply calculated by expanding the exciton wavefunction in the electron–hole basis and calculating the strain field associated with a single longitudinal acoustic (LA)

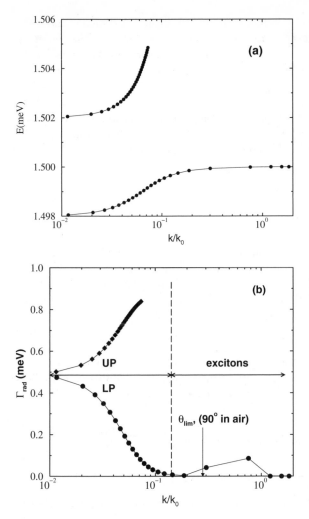

Fig. 6.1. (a) Exciton polariton dispersion at small k. (b) Radiative recombination rates. The *symbols* mark the points used for discretization, evenly spaced in energy. The *lines* are only guides for the eye. The radiative rates of excitons having $k > k_0 = n\omega/c$ are zero

phonon. This results in the following interaction Hamiltonian [180]:

$$\mathcal{H}_{\text{exc-ph}} = \sum_{\boldsymbol{k},\boldsymbol{q},q_z} \sqrt{\frac{\hbar |\boldsymbol{q}|}{\rho u \mathcal{V}}} \left[a_{\text{e}} I^{\parallel} \left(\frac{m_{\text{h}}}{M} |\boldsymbol{q}| \right) I_{\text{e}}^{\perp}(q_z) - a_{\text{h}} I^{\parallel} \left(\frac{m_{\text{e}}}{M} |\boldsymbol{q}| \right) I_h^{\perp}(q_z) \right]$$

$$\times \left(b_{\boldsymbol{k}+\boldsymbol{q}}^{\dagger} b_{\boldsymbol{k}} a_{\boldsymbol{q},q_z} + \text{h.c.} \right). \tag{6.4}$$

Here $\mathcal{V}$ is the quantization volume, u the speed of the LA phonon, and ρ the mass density of the solid. $I^{\parallel}(q)$ and $I^{\perp}_{\mathrm{e\,(h)}}(q_z)$ are the superposition integrals of the exciton in-plane wavefunctions and the phonon waves. We consider the following exciton envelope wavefunction:

$$\Psi_k(\boldsymbol{\rho}, z_{\mathrm{e}}, z_{\mathrm{h}}) = F_k(\boldsymbol{\rho})c(z_{\mathrm{e}})v(z_{\mathrm{h}}) , \tag{6.5}$$

where $F_k(\rho) = \sqrt{2/\pi a_{\mathrm{B}}^2}\exp(-\rho/a_{\mathrm{B}})$ is the in-plane exciton envelope function, c and v are the electron and hole confinement functions, and a_{B} is the exciton Bohr radius. Then, the superposition integrals read [180]

$$I^{\parallel}(|\boldsymbol{q}|) = \left[1 + \left(\frac{qa_B}{2}\right)^2\right]^{-3/2} ,$$

$$I^{\perp}_{\mathrm{e}}(q_z) = \int \mathrm{d}z |c(z)|^2 \mathrm{e}^{\mathrm{i}q_z z} . \tag{6.6}$$

Expressions analogous to (6.6) hold for $I^{\perp}_{\mathrm{h}}$, with m_{e} in place of m_{h} and $v(z)$ in place of $c(z)$. These integrals introduce cutoffs in the scattering from phonons having $q_{\parallel} > 1/a_{\mathrm{B}}$ or $q_z > 2\pi/L$ (L is the QW width). For infinite well confinement potentials,

$$I^{\perp} = \frac{1}{q_z L/2} \frac{\sin(q_z L/2)}{1 - [q_z L/(2\pi)]^2}. \tag{6.7}$$

The scattering of exciton polaritons from phonons originates from the exciton component only and thus the scattering rate contains the Hopfield factors of (1.37) squared. The rates can be calculated with the Fermi golden rule:

$$W^{\pm}_{i,\boldsymbol{k}\to j,\boldsymbol{k}'} = \frac{2\pi}{\hbar}\sum_{\boldsymbol{q},q_z}\left|X^{\star}_{j,\boldsymbol{k}'}\langle \boldsymbol{k}'|\langle 0_{\boldsymbol{q},q_z}|H_{\mathrm{exc-ph}}|1_{\boldsymbol{q},q_z}\rangle|\boldsymbol{k}\rangle X_{i,\boldsymbol{k}}\right|^2$$

$$\times \delta\left(\hbar\Omega_{j,\boldsymbol{k}'} - \hbar\Omega_{i,\boldsymbol{k}} \pm E^{(ph)}_{\boldsymbol{q},q_z}\right)\left[\dot{N}(E^{(ph)}_{\boldsymbol{q},q_z}) + 1/2 \pm 1/2\right] . \tag{6.8}$$

Here the $+$sign is for phonon emission and the $-$sign for absorption. Integration of the delta function yields

$$W^{\pm}_{i,\boldsymbol{k}\to j,\boldsymbol{k}'} = \frac{4\pi}{\hbar}\frac{L_z}{2\pi}\frac{\hbar}{2\rho\mathcal{V}u}\frac{\left(|\boldsymbol{k}-\boldsymbol{k}'|^2 + q_z^{(0)^2}\right)}{|\hbar u q_z^{(0)}|}|X^{\star}_{j,\boldsymbol{k}'}X_{i,\boldsymbol{k}}|^2$$

$$\times \left[a_{\mathrm{e}}I^{\parallel}\left(\frac{m_{\mathrm{h}}}{M}|\boldsymbol{k}'-\boldsymbol{k}|\right)I^{\perp}_{\mathrm{e}}(q_z^{(0)}) - a_{\mathrm{h}}I^{\parallel}\left(\frac{m_{\mathrm{e}}}{M}|\boldsymbol{k}'-\boldsymbol{k}|\right)I^{\perp}_{\mathrm{h}}(q_z^{(0)})\right]^2$$

$$\times \left[N(E^{(ph)}_{\boldsymbol{k}'-\boldsymbol{k},q_z^{(0)}}) + 1/2 \pm 1/2\right] , \tag{6.9}$$

where $q_z^{(0)}$ is

$$q_z^{(0)} = \left(\left(\frac{E_{j,\boldsymbol{k}'} - E_{i,\boldsymbol{k}}}{u}\right)^2 - |\boldsymbol{k}-\boldsymbol{k}'|^2\right)^{1/2} . \tag{6.10}$$

In order to derive the total rate of scattering from the energy bin i_1 to the energy bin i_2, we have to integrate over all possible angles θ between $\boldsymbol{k}$ and $\boldsymbol{k}'$ and over all the possible final states in the interval $[E_{i_2} - \Delta E/2, E_{i_2} + \Delta E/2]$. Assuming that the scattering rate is only mildly dependent on this final energy (see the discussion above), this integration simply yields a factor $\Delta E/2\partial_{k^2} E(k_2)$. Finally, we obtain

$$W_{i_1,i_2} = \frac{2\pi}{\hbar} \frac{S}{(2\pi)^3} \frac{\Delta E}{2\partial_{k^2} E(k_2)} X_{k_1} X_{k_2} \frac{\hbar \Delta k}{2\rho S u} R'(k_1, k_2)$$
$$\times [N_{\mathrm{ph}}(|E_2 - E_1|) + 1/2 \pm 1/2] \tag{6.11}$$

$$R'(k_1, k_2)$$
$$= 2 \int_0^{\theta_{\max}} \mathrm{d}\theta \frac{2\Delta k}{\hbar u q_z} I_\perp^2(q_z) \left[a_e I_\parallel \left(\beta \Delta k_\parallel a_B \right) - a_h I_\parallel \left(\alpha \Delta k_\parallel a_B \right) \right]^2, \tag{6.12}$$

where $q_z(\theta) = \sqrt{\Delta k^2 - \Delta k_\parallel^2}$, $\Delta k_\parallel^2(\theta) = k_1^2 + k_2^2 - 2k_1 k_2 \cos\theta$, $\Delta k = |E_2 - E_1|/\hbar u$, and

$$\cos\theta_{\max} = \begin{cases} 1, & c > 1 \\ c, & c \in [-1,1] \\ -1, & c < -1 \end{cases} , \quad c = \frac{k_1^2 + k_2^2 - \Delta k^2}{2k_1 k_2}.$$

In the numerical calculation of these rates, we used adaptive integration for the angular integration over θ, after changing the variable to smooth the singularity at the extremum $\theta_{\max}$.

Typically, the exciton–phonon scattering can be characterized by the total out-scattering rate of the $\boldsymbol{k} = 0$ exciton to larger-$k_\parallel$ excitons by phonon absorption. Above a few K, this rate becomes linear in temperature:

$$\sum_k W_{k \to k} = \gamma T. \tag{6.13}$$

Experimentally, the rate $\gamma = 5\,\mu\mathrm{eV/K}$ for a 100 Å GaAs QW. This same rate can be obtained from the above calculation by using deformation potentials $a_e - a_h \sim 13$ eV, $m_e = 0.07\,m_0$, $m_h = 0.17\,m_0$, and $a_B = 90$ Å. Thus, the cutoff energy for phonon absorption and emission is typically

$$E_{\mathrm{cut}} = \hbar u \frac{2\pi}{100\,\text{Å}} \sim 1\ \mathrm{meV}.$$

When the QW is embedded in a cavity, and strong coupling sets in, the $\boldsymbol{k} = 0$ exciton becomes mixed with the cavity photon, resulting in the Rabi-split upper and lower polariton. Phonon absorption from the $\boldsymbol{k} = 0$ lower polariton causes scattering either into other lower polaritons or into exciton-like polaritons, at an energy $\Omega/2$ above. The scattering into other lower polaritons is greatly reduced with respect to the rate in (6.13), by a factor $m_{\mathrm{pol}}/m_{\mathrm{exc}}$, owing to the reduction of the density of states (DOS). This

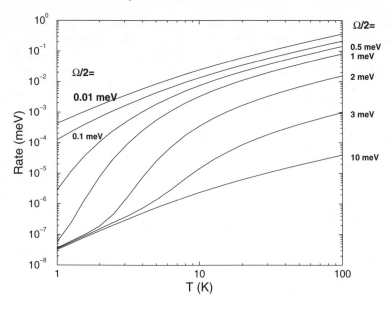

Fig. 6.2. Acoustic-phonon absorption rate of $k = 0$ lower polaritons as a function of the lattice temperature, for different Rabi splittings Ω. Other parameters as in the text

amounts to almost four orders of magnitude. When $\Omega/2 > E_{\text{cut}}$, phonon absorption causing scattering into the exciton-like polaritons is greatly reduced when $k_{\text{B}}T < \Omega/2$, and scattering to the exciton-like polaritons becomes active for larger temperatures only. We plot in Fig. 6.2 the total phonon absorption rate of the $k = 0$ exciton polariton as a function of the lattice temperature for different Rabi splittings. The suppression of phonon scattering for $\Omega/2 > 1$ meV is evident. For smaller Ω, the reduction of the DOS, even at $k = 0$, is less marked, and the resulting phonon absorption rate is larger. We may also notice the onset of absorption leading to exciton-like polaritons for $k_{\text{B}}T > \Omega/2$ at intermediate splitting, and the extended suppression of phonon scattering for the largest splitting considered. This because $\Omega \gg 1$ meV in this case, and even the phonon scattering to exciton-like polaritons is somewhat suppressed.

For the upper polariton, the situation is different. The upper polariton is degenerate in energy with exciton-like polaritons. Thus, both phonon absorption and phonon emission events to the exciton-like branch are possible from the $k = 0$ upper polariton, and the total out-scattering rate is dominated by these interbranch scattering events. Moreover, the scattering rate is of the same order of magnitude as the exciton scattering rate introduced above in (6.13), and thus is orders of magnitude larger than the lower-polariton scattering rate for large enough Rabi splittings.

In the previous discussion we did not consider the interaction of excitons with longitudinal optic phonons. The strongest interaction of this type in the polar III–V materials is electrostatic and results in the Fröhlich interaction. Typical LO phonon energies range from 20 to 40 meV in III–V materials, and are usually larger that the exciton binding energy. LO phonon absorption is thus a ionizing event. The resulting dynamics therefore involves free carriers, and is outside our scope. In GaAs $\hbar\omega_{LO} = 36$ meV, and thus rather large temperatures are needed to activate LO phonon absorption. Actually, this interaction is much stronger than the LA phonon interaction, and thus the two phonon absorption rate become comparable at a temperature of about 40 K. In the following, we shall always take $T < 40$ K, so that LO phonon absorption precesses are negligible. Of course, LO phonon absorption and acoustic-phonon absorption from lower polaritons can become comparable at smaller temperatures, since the latter rate is suppressed as explained before. However, both scattering rates are basically negligible, when compared with the fast radiative recombination of these modes.

6.2.3 Exciton–Exciton Scattering

We introduced the exciton concept in Sect. 1.2.1 and remarked that it represents a bosonic excitation at low densities, when $n_{exc}a_B^2 \ll 1$. Let us explain this argument in more detail. Consider a state of the crystal with two excitons: this state approximately diagonalizes the total Hamiltonian, $K_e + K_h + V_{eh} + V_{ee} + V_{hh}$, introduced in (1.11). The remainder consists of terms of the order a_B^2/S, S being the sample surface area. Iterating this argument N times, we find that a state with N excitons is an approximate eigenstate of the total Hamiltonian apart from terms of order $Na_B^2/S = n_{exc}a_B^2$. The concept of composite bosons is very valuable. It is, of course, in principle possible to describe any state of the crystal as a combination of electron and hole states, but in certain conditions this is awkward. Consider, for example, an excitonic condensate or, more generally, any nonequilibrium state of the crystal where a large number of excitons populate the same state. The description of an exciton as an *ideal* boson is, however, seldom sufficient in the dynamics, except at very low densities. The excess terms of order $n_{exc}a_B^2$ discussed above do have in fact a physical significance, and represent an additional energy related to the finite overlap of the fermionic constituents of the excitons. We thus understand that it is possible to treat the excitons as *interacting* bosons, choosing an interaction Hamiltonian between the bosons such that its expectation value for the N-exciton state approximately reproduces the excess terms discussed above, which are in fact generated in the fermionic space. This procedure can be formalized, by mapping the fundamental microscopic (fermionic) Hamiltonian into a bosonic Hamiltonian that contains the exciton kinetic-energy term as the lowest-order term, and contains higher-order terms representing two-body, three-body, etc., exciton interactions. There are systematic methods for constructing the higher-order terms, which include, among many others commutator expansions [181] and ordering/transcription operator methods [182, 183]. In practice, we have to use a truncated bosonic Hamiltonian. The simplest contains only the excitonic kinetic terms, and represents ideal bosons. The next order includes two-body exciton–exciton interactions. In the calculation of dynamical effects, we shall use this truncated Hamiltonian. However, the calculated quantities are non-perturbative in the interaction terms, and, in principle, what has been left out of the microscopic Hamiltonian because of the truncation could introduce nonnegligible contributions to the final results. We are not aware of a mathematical treatment of this problem in this context, and we can only resort to a phenomenological justification, that the two-body interaction is satisfactory for describing many aspects of the dynamics of real systems. In the case of thermal equilibrium, the two-body interaction terms have to be used, and are sufficient, in the description of second-order phase transitions [184], such as the onset of superfluidity. In contrast, we know a priori that the phase transition to the electron–hole plasma is not at all described in the bosonic model, even when two-body interaction terms are introduced. In conclusion, two-body interaction terms are expected to produce a good description of the dynamics of our system provided that this is restricted to the excitonic phase, with $n_{\mathrm{exc}} a_{\mathrm{B}}^2 < 1$.

The exciton–exciton interaction Hamiltonian reads

$$H_{\mathrm{exc-exc}} = \frac{1}{4} \sum_{k,k',q} M_{\boldsymbol{k}_1,\boldsymbol{k}_2,\boldsymbol{q}} b^\dagger_{\boldsymbol{k}_1+\boldsymbol{q}} b^\dagger_{\boldsymbol{k}_2-\boldsymbol{q}} b_{\boldsymbol{k}_1} b_{\boldsymbol{k}_2}, \tag{6.14}$$

where we have omitted spin degrees of freedom. The calculation of $M_{\boldsymbol{k}_1,\boldsymbol{k}_2,\boldsymbol{q}}$ has already been carried out in the literature, including the spin degrees of freedom [183, 185, 186]. In GaAs quantum wells, for heavy-hole excitons, we have four exciton spin states, of which two are dark. Rapid scattering between the different spin states equalizes their population. We shall not address this equalization dynamics, but shall always assume equal population among all the spin states. Averaging out the spin degrees of freedom results in a factor $1/2$ in the calculation of the scattering rates, with respect to the case of aligned spins. Both the exchange of constituents and the direct Coulomb interaction contribute to $M_{\boldsymbol{k}_1,\boldsymbol{k}_2,\boldsymbol{q}}$. However, the direct Coulomb interaction (of dipole–dipole type) is much weaker than the short-range exchange, and here we neglect it completely. For small momenta $k, k', q \ll a_{\mathrm{B}}^{-1}$,

$$M \sim 2 \sum_{k,k'} V_{\boldsymbol{k}-\boldsymbol{k}'} \phi_{\boldsymbol{k}} \phi_{\boldsymbol{k}'} \left(\phi_{\boldsymbol{k}}^2 - \phi_{\boldsymbol{k}} \phi_{\boldsymbol{k}'} \right) \sim 6 E_{\mathrm{B}} \frac{a_{\mathrm{B}}^2}{S}. \tag{6.15}$$

Here $V_{\boldsymbol{k}} = 2\pi/(Sk)$ is the two-dimensional Coulomb interaction, and $\phi_{\boldsymbol{k}} = \sqrt{8\pi/S}[1 + (ka_{\mathrm{B}})^2]^{-3/2}$ is the 1s two-dimensional exciton wavefunction. The detailed momentum and angular dependence of $M_{\boldsymbol{k},\boldsymbol{k}',\boldsymbol{q}}$ has been calculated numerically in [186, 187]. This dependence typically shows a momentum cutoff of the order of a_{B}^{-1}, which we neglect as we shall consider only cold distributions of excitons spanning a smaller phase space.

The bosonization of the interaction Hamiltonian between electrons, holes, and photons is also straightforward, and gives linear and nonlinear terms. The linear interaction term has been discussed in detail in Sect. 2.1.2, and represents the creation and annihilation of excitons from and into photons. The higher-order nonlinear term is

$$H^{(4)}_{\text{eh-phot}} = \frac{1}{2} \sum_{\boldsymbol{k}_1, \boldsymbol{k}_2, \boldsymbol{q}} \sigma_{\boldsymbol{k}_1, \boldsymbol{k}_2, \boldsymbol{q}} \left(b^\dagger_{\boldsymbol{k}_1 + \boldsymbol{k}_2 - \boldsymbol{q}} b_{\boldsymbol{k}_1} b_{\boldsymbol{k}_2} + \text{h.c.} \right) \left(a_{\boldsymbol{q}} + a^\dagger_{-\boldsymbol{q}} \right), \quad (6.16)$$

and represents the scattering of an exciton by another exciton created by absorption of a photon. The matrix element $\sigma_{\boldsymbol{k}_1, \boldsymbol{k}_2, \boldsymbol{q}}$ features the usual cut-offs of order a_B^{-1}. In the limit of $k_1, k_2, q \ll a_B^{-1}$,

$$\sigma \sim \frac{4\pi}{7} \frac{a_B^2}{S} \Omega.$$

This nonlinear exciton–photon interaction term is the bosonic counterpart of the phase-space-filling effect in the interaction of electrons and holes with photons, i.e. absorption of photons is blocked by filling of the fermionic states. In the microcavities that we are considering, $\Omega \sim 4$ meV, while $E_B \sim 10$ meV. In the calculation of the scattering the matrix elements are squared. This non-linear interaction gives a contribution of order $(\Omega/4E_B)^2 \sim 10^{-2}$ compared with exciton–exciton scattering, and is fairly negligible [188]. Thus, phase space filling is negligible in a microcavity polariton system. This conclusion is, however, related to the assumption of full translational symmetry, and will be discussed in more detail later.

We now calculate $Y_{ij,i_1 j_1}$ of (6.3) using the Fermi golden rule. The calculation of the elastic exciton–exciton scattering rate follows that of Snoke and Wolfe [189], where the energy conservation delta function is integrated using

$$\int d\boldsymbol{k}_2 d\boldsymbol{q} \, \delta(E_1 + E_2 - E_3 - E_4)$$

$$= \int d\boldsymbol{k}_2 \, d\boldsymbol{q} \frac{1}{\partial_{k^2} E(k_4)} \delta[k_4^2 - (k_2^2 + q^2 - 2k_2 q \cos\theta_{\widehat{k_2 q}})]$$

$$= \int dk_2^2 \int_0^\pi d\theta_{\widehat{k_2 q}} d\boldsymbol{q} \frac{\delta(\theta_{\widehat{k_2 q}} - \theta_0)}{\partial_{k^2} E(k_4) 2 k_2 q |\sin\theta_0|},$$

where $\cos\theta_0 = [k_4^2 - (k_2^2 + q^2)]/(2k_2 q)$, and $k_4 = k(E_4 = E_1 + E_2 - E_3)$ is then fixed by energy conservation. The integration over $d\boldsymbol{q}$ is then carried out using

$$\int d\boldsymbol{q} = \int dq^2 \int_0^\pi d\theta_{\widehat{q k_1}} = \int dq^2 \frac{dk_3^2}{2 k_1 q |\sin\theta_{\widehat{q k_1}}|},$$

where $\cos\theta_{\widehat{q k_1}} = [k_3^2 - (k_1^2 + q^2)]/(2k_1 q)$. Finally, using also $dk_2^2 = dE_2/[2\partial_{k^2} E(k_2)]$, $dk_3^2 = dE_3/[2\partial_{k^2} E(k_3)]$, and $x = q^2$, we obtain

$$Y_{i_1i_2,i_3i_4}=\frac{\pi}{\hbar}\frac{S^2}{(2\pi)^4}\frac{\Delta E^2}{\partial_{k^2}E(k_2)\partial_{k^2}E(k_3)}|X_{k_1}X_{k_2}X_{k_3}X_{k_4}|^2\,|M|^2$$
$$\times\frac{R(k_1,k_2,k_3,k_4)}{\partial_{k^2}E(k_4)}, \tag{6.17}$$

$$R(k_1,k_2,k_3,k_4)=\frac{1}{2}\int_{\mathcal{I}}\mathrm{d}x\frac{1}{\sqrt{[(k_1+k_3)^2-x][x-(k_1-k_3)^2]}}$$
$$\times\frac{1}{\sqrt{[(k_2+k_4)^2-x][x-(k_2-k_4)^2]}}. \tag{6.18}$$

The existence of the cosines above imposes the limits of integration:

$$\mathcal{I}=[(k_1-k_3)^2,(k_1+k_3)^2]\cap[(k_2-k_4)^2,(k_2+k_4)^2].$$

The $|X_k|^2$ are the squared Hopfield coefficients for the exciton introduced in Sect. 2.1.2 and specify the exciton content of the polariton state. We used adaptive integration to evaluate (6.18), after changing the variable to smooth the singularity at the integration extrema.

The exciton–exciton scattering rate can be characterized by the scattering out from the $k=0$ exciton state resulting from hot excitons. This scattering rate depends mildly on the exciton population distribution n_k, in that it is smooth compared with ΔE and has an extent less than a_{B}^{-1}. Considering a thermal distribution of excitons, this amounts to considering $\hbar^2 a_{\mathrm{B}}^{-2}/2m_{\mathrm{exc}} > k_{\mathrm{B}}T > \Delta E$. For a GaAs QW, using $\Delta E =0.05$ meV, this means 0.5 K $< T <$ 30 K. Above 30 K the detailed form of the scattering matrix element $M_{k,k',q}$ needs to be considered and included in the calculation of the scattering rates, which is straightforward. Qualitatively, this simply produces a decrease of the scattering rates. The scattering out from the $k=0$ exciton state has already been described in (6.2), and is simply linear in the density. When we consider strong coupling, the $k=0$ exciton becomes mixed with the photon, resulting in two new modes, the lower and upper polaritons. We may still, however, consider that only the exciton-like branch is thermally populated, as shown in the next section. We then find, in analogy with the phonon-scattering case discussed in the previous section, that upper polaritons are scattered to the exciton-like branch, and that intraband scattering is fairly negligible owing to the reduced density of states. Lower-polariton scattering ($k=0$) is suppressed for large enough splittings. The overall behavior is very similar to that presented in Fig. 6.2 for phonons. We show the scattering rate for the $k=0$ lower polariton as a function of the exciton temperature, for a fixed exciton density of 10^9 cm^{-2} and for different Rabi splittings, in Fig. 6.3.

6.3 Integration of the Rate Equation: Results

We have integrated numerically the rate equations (6.3). We considered a resonant pump term of Gaussian type, centered at the exciton energy $E=0$,

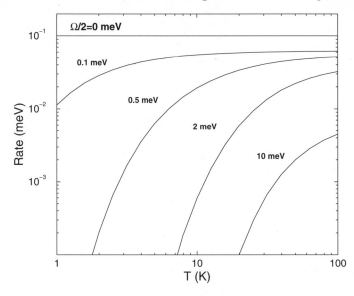

Fig. 6.3. Scattering rate of the $k = 0$ lower polaritons as a function of the exciton-like-polariton reservoir temperature, at a fixed density of $n_{\mathrm{exc}} = 10^{10}$ cm^{-2}, for different Rabi splittings Ω. Other parameters as in the text

and different lattice temperatures and excitation intensities. The pump term reads

$$P_i = Pe^{-E(k_i)^2/\sigma^2}. \tag{6.19}$$

We chose $\sigma = 0.25$ meV, representative of a quasi-monochromatic pump beam, larger than $\Delta E = 0.05$ meV, but smaller than E_{B}. In Fig. 6.4a we show the exciton polariton populations versus energy, for different pump intensities at $T = 4$ K. The resulting total density is also shown in the figure, and is essentially determined by the averaged radiative recombination rate of the excitons [190]. At low densities, the exciton-like polaritons of the lower branch with $E > 0$ are quasi-thermally populated, showing a Boltzmann factor on the figure, i.e. a linear tail on the logarithmic scale. Both upper polaritons and lower polaritons with $E < 0$ are strongly depleted with respect to this thermal population. This is a bottleneck effect in the thermalization of the exciton-like polaritons into the lower polaritons. It is discussed in detail in [191]. This effect is a result of the combination of a strong reduction of the exciton–phonon scattering rates to the small-k lower polaritons discussed above, and the short radiative lifetimes of these states. In the figure, we may also notice, that on increasing the density, exciton–exciton scattering becomes relevant. It results in a sizeable filling of the depleted LP and UP populations: the slope of the LP population curve for $E < 0$ becomes smaller. Also, the small residual dip [180] in the population distribution related to the radiative excitons close to $E = 0$ fills up, disappearing at densities around

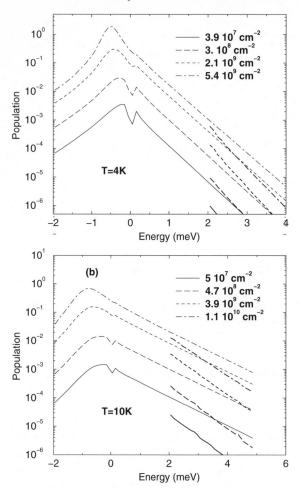

Fig. 6.4. Exciton and LP populations versus energy at different densities, at (**a**) $T = 4$ K, (**b**) $T = 10$ K. The UP population is marked by the *thicker lines*, starting at $E = 2$ meV

10^9 cm^{-2}. In Fig. 6.4b we show the corresponding population distributions at $T = 10$ K. The same pump densities were used to generate these data. We remark that larger densities are obtained, given the same pump rates, at $T = 10$ K than at 4 K. This is related to the thermally averaged radiative lifetime of the excitons, which increases with temperature [190]. Otherwise, we observe the same qualitative behavior as described above for $T = 4$ K. Because of faster phonon scattering, the depletion of the LPs and UPs is slightly less marked at $T = 10$ K than at $T = 4$ K.

In the energy interval -0.5 meV $< E < 0$ meV, the lower polaritons have a dominantly exciton-like character, and a DOS smaller than that of excitons, but larger than that of $k = 0$ polaritons. They also have longer radiative lifetimes than at $k = 0$, as is clearly shown in Fig. 6.1b. Exciton–phonon scattering is thus somewhat more efficient than at $k = 0$, and radiative recombination is slower: both effects finally result in an occupation factor larger than one in this region of energies, at a density $n_{exc} \sim 5 \times 10^9$ cm^{-2}. This is indicative of the possible onset of quantum statistical effects, namely stimulated scattering, and eventually the onset of superfluidity. The latter effect is well beyond our rate-equation approach, and needs more sophisticated techniques to explore it, techniques which take into account the renormalization of the quasiparticle energies corresponding to the large occupations found. However, this result is suggestive of interesting dynamics that could develop in this transition region at intermediate densities.

6.4 Interface Disorder and Bottleneck Dynamics

The currently grown QW samples have QW interfaces which are far from ideal. The interface disorder potential is not well characterized, but in the best samples it presumably results from monolayer fluctuations, with correlation lengths of the order of a few lattice spacings. Weakly bound excitons in III–V materials extend over many thousands of lattice sites and effectively average this potential to a fluctuation of a few meV in amplitude and with a correlation length of the order of a_B. This potential breaks the in-plane translational symmetry, and results in an inhomogeneous broadening of the excitonic absorption. This inhomogeneous linewidth does not show all of the fluctuation amplitude of disorder potential, because the exciton has a finite mass. Confining the exciton to a single in-plane potential well costs kinetic energy; thus, the exciton prefers to spread out, and this further averages the fluctuations of the disorder potential. This effect is called "motional narrowing". Polaritons close to $k = 0$ have an effective mass which is orders of magnitude smaller than the exciton mass. Thus, motional-narrowing effects should be more prominent for polaritons close to $k = 0$. This was suggested by Whittaker et al. in [192]. However, the situation in the current samples is more complex, owing to the existence of exciton-like polaritons and to the large strength of the disorder potential. This potential is strong enough to substantially mix polaritons close to $k = 0$ and exciton-like polaritons. Thus, even the eigenstates of the system having eigenenergies close to $\pm\Omega/2$, which are remnants of the $k = 0$ polaritons, have a strong component of exciton-like polaritons. Motional narrowing is strongly reduced with respect to the naively expected four orders of magnitude [193, 194]. Moreover, as upper polaritons are degenerate in energy with $E > 0$ exciton-like lower polaritons, motional narrowing for them is substantially similar to that found for the excitons. This difference between lower and upper polaritons follows just the same trend as that remarked on above for phonon and exciton–exciton scattering. In summary, mixing with a cavity photon does not result in the large motional-narrowing effects that are expected for a weakly disordered system. It is also worth noticing that the absorption lineshapes are well reproduced with a simplified model which accounts for the effects of disorder at the level of the exciton absorption lineshape only [195], again suggesting no role of a reduced polariton mass arising from the exciton–cavity photon mixing in the determination of the inhomogeneous lineshape of the polariton, except for the introduction of an energy shift, the Rabi splitting.

Both of the models of the inhomogeneous lineshape suggest that a simple "clean" polaritonic picture of the dynamics is inadequate for current samples. Certainly, the strong disorder affects the dynamics in a nonperturbative fashion; we remarked above that it qualitatively changes the nature of the eigenstates of the system. Scattering of exciton polaritons on disorder cannot be treated perturbatively and, in particular, it cannot be directly inserted into the kinetic approach used in this work. The results obtained in our "clean" polariton picture thus have to be discussed in connection with this observation. As can be expected, some predictions of the polariton dynamics in our "clean" polariton picture are in striking contrast to the experimental observations. Owing to the bottleneck effect, and the favored population of the $k = 0$ upper polaritons with respect to the lower polaritons, the photoluminescence lineshape is expected to be fairly nonthermal in our "clean" polariton model. However, this is not the case in the experiments. Some of these experiments involve excitation of the free carriers, and thus involve the additional problems discussed at the beginning of this chapter. We shall not go into more detail about these experiments. However, even resonant-like excitation at $k = 0$ and $E = 0$ failed to show the expected strong deviations from thermalization [196]. We also carried out similar experiments, but directly injecting larger exciton-like populations at finite k (see later). However, even in this case, lower-polariton emission dominates over upper-polariton emission at low temperatures, suggesting a much faster phonon emission rate to lower polaritons than that expected from the reduced-DOS argument. Nevertheless, this does not mean that the bottleneck effect is completely lifted. Indeed, none of these experiments establishes a quantitative ratio between the lower-polariton density close to $k = 0$ and that of exciton-like polaritons. In our experiments, estimating the exciton density and the absolute lower-polariton emission rates, we could qualitatively conclude that the bottleneck effect is indeed present. We notice that indirect, but solid, experimental evidence of the bottleneck effect in the relaxation dynamics has already been established in the boser experiments presented in [216] and carried out independently by many groups [197, 198]. If the exciton-like polaritons could relax and thermalize into lower polaritons at $k = 0$, the quasi-condensation of these lightweight particles would occur at much lower densities than those where strong coupling collapses. Moreover, the bottleneck effect implies that the exciton dynamics of bare and cavity-embedded QWs are similar, and independent of the exciton cavity detuning [199], as observed experimentally by measuring photoluminescence decay times [200].

Finally, we notice that the bosonization approach itself is not applicable when the disorder is strong enough to completely localize the excitons. The exciton diffusion lengths are then reduced to just a_B, and the exciton behaves as a two-level system, with fundamental fermionic character. Mixing with cavity photons may partially recover the bosonic properties of lower polaritons close to $k = 0$. The lower-polariton localization length is in fact necessarily larger than that of the bare exciton, because of the light photon mass. This is true despite the fact that we know from the above discussion that this effect is not as large as expected. Thus, many quasi-degenerate exciton states participate in a given delocalized lower-polariton state, and the polariton occupation numbers can be larger than one again. The actual maximum number of polaritons allowed in a mode depends on the details of the polariton eigenstate in the disorder potential. Saturation effects at small occupation can then be described as effective polariton–polariton scatterings. We remark that in this case, the main nonlinearity is related to phase space filling, whereas exciton–exciton scattering is obviously suppressed by localization. This is the opposite of what we expected in the "clean" polariton case. Four-wave-mixing experiments in currently available samples have confirmed that the main nonlinearity term is the phase-space-filling term [201], thus pointing to a strong localization effect. However, the same experiments established a finite amount of exciton–exciton scattering, pointing to an intermediate situation between the one described by our clean model and the one discussed above for completely localized excitons. In summary, although we know that the results of our model are quantitatively inapplicable to the currently grown systems, we understand that they are qualitatively sound. In particular, there are linear dynamic effects mediated by phonons, nonlinear effects mediated by two-body scattering, and the possibility of large lower-polariton populations due to a partial delocalization.

6.5 Stimulated Scattering of Lower Polaritons: Theory

The bosonicity of the small-k lower polaritons is central to the physics of the microcavity system and to its potential applications. The possibility of obtaining stimulated emission relies on this fundamental property. In this section we present preliminary explorations which are aimed at establishing this important point and clarifying the dynamics of the system in greater detail. From the discussion in the previous section and the results of the simulations within our model, we conclude that the existence of the bottleneck in the relaxation of excitons to LPs is a solid assumption. It follows that in actual experiments the small-k lower-polariton population ("LP population" in the following) and the exciton-like polariton population (simply "excitons" in the following) can be independently controlled by different optical sources.

This opens the way to performing scattering experiments between LPs and excitons. In particular, we can identify five main processes:

1   exciton $\rightarrow$ LP + phonon
2   exciton $\rightarrow$ UP$\pm$ phonon
3   exciton + exciton $\rightarrow$ LP + exciton
4   exciton + exciton $\rightarrow$ UP + exciton
5   exciton + exciton$\rightarrow$ LP + UP.

Processes 1, 3, and 5 are stimulated by large LP populations. Interestingly, the scattering rate of process 5 may be measured from the UP emission rate. This allows experimental access to the direct measurement of the actual LP population through the observation of stimulated scattering.

We showed in Sect. 6.3 that the excitons are basically thermalized. Here we simplify the rate equations, considering the excitons as a thermal reservoir with fixed temperature. We then consider large LP populations at $\boldsymbol{k} = 0$, injected from an external pump. Otherwise, the LPs and UPs are negligibly populated. We therefore neglect those scatterings involving LPs or UPs which are not stimulated, for the usual reason that the final DOS and the scattering rate are negligible in these cases. For the same reason, the out-scattering rates from the LPs and UPs have been neglected, compared with the fast radiative-recombination terms. We then obtain an overall equation for the total exciton population N_{exc}:

$$\frac{dN_{\mathrm{exc}}}{dt} \tag{6.20}$$

$$= P_{\mathrm{exc}} - \frac{N_{\mathrm{exc}}}{\tau_{\mathrm{exc}}} - \left(a_{\mathrm{LP},\boldsymbol{k}=0} n_{\mathrm{exc}} + \sum_{\boldsymbol{k}} b'_{\boldsymbol{k}} n_{\mathrm{exc}}^2 + b_{\mathrm{LP},\boldsymbol{k}=0} n_{\mathrm{exc}}^2 \right) N_{\mathrm{LP}}.$$

We also obtain equations for the UPs, $N_{\mathrm{UP},\boldsymbol{k}}$ and for the LPs at $\boldsymbol{k} = 0$, N_{LP}:

$$\frac{dN_{\mathrm{UP},\boldsymbol{k}}}{dt} = -\frac{N_{\mathrm{UP},\boldsymbol{k}}}{\tau_{\mathrm{UP},\boldsymbol{k}}} + a_{\mathrm{UP},\boldsymbol{k}} n_{\mathrm{exc}} + [b_{\mathrm{UP},\boldsymbol{k}} + b'_{\boldsymbol{k}}(1 + N_{\mathrm{LP}})] n_{\mathrm{exc}}^2, \tag{6.21}$$

$$\frac{dN_{\mathrm{LP}}}{dt} = P_{\mathrm{LP}} - \frac{N_{\mathrm{LP}}}{\tau_{\mathrm{LP}}} + a_{\mathrm{LP},\boldsymbol{k}=0}\, n_{\mathrm{exc}}(1 + N_{\mathrm{LP}})$$

$$+ \left(b_{\mathrm{LP},\boldsymbol{k}=0} + \sum_{\boldsymbol{k}} b'_{\boldsymbol{k}} \right) n_{\mathrm{exc}}^2 (1 + N_{\mathrm{LP}}). \tag{6.22}$$

Here $n_{\mathrm{exc}} = N_{\mathrm{exc}}/S$ is the exciton density, $\tau_{\mathrm{LP}} = (\Gamma_{\boldsymbol{k}=0}^{(1)})^{-1}$ is the LP lifetime, $\tau_{\mathrm{UP},\boldsymbol{k}}$ is the UP radiative lifetime, and τ_{exc} is the exciton radiative lifetime averaged over a thermal distribution [190].

The coefficients a and b can be calculated as before from the Fermi golden rule:

$$a_{j,k_2} = \frac{2\pi}{\hbar} \frac{S}{(2\pi)^2} \int \frac{dE_1}{2\partial_{k^2} E(k_1)} |X_{k_1}^{(1)} X_{k_2}^{(j)}|^2 \frac{\hbar \Delta k}{2\rho S u}$$

$$\times R'(k_1, k_2) N_{\mathrm{ph}}(E_2 - E_1) f_{k_1}, \tag{6.23}$$

where R', N_{ph}, Δk, and all other quantities used here were introduced in (6.12). We have also explicitly included the branch index $j = 1$. The distribution f_k is the normalized Boltzmann distribution of the exciton reservoir.

The quadratic coefficients are given by

$$b_{j,\boldsymbol{k}_3} = \frac{\pi}{\hbar} \frac{S^2}{(2\pi)^4} \int \mathrm{d}k_1^2 \mathrm{d}k_2^2 \, |X_{k_1}^{(1)} X_{k_2}^{(1)} X_{k_3}^{(j)} X_{k_4}^{(1)}|^2 \, |M|^2$$

$$\times \frac{R(k_1, k_2, k_3, k_4)}{\partial_{k^2} E^{(1)}(k_4)} f_{\boldsymbol{k}_1} f_{\boldsymbol{k}_2}. \tag{6.24}$$

Owing to the small LP and UP DOSs, the major contribution originates from excitons as final scattering states in $\boldsymbol{k}_4$. This rate can be analytically calculated [188], and results in:

$$b_{j,\boldsymbol{k}_3} \sim \pi |X_{k_3}^{(j)}|^2 |M|^2 2 m_{\mathrm{exc}} \frac{S^2}{(2\pi)^4}$$

$$\times \int_{k_1^2 k_2^2 > m_{\mathrm{exc}} E_3} \mathrm{d}k_1^2 \mathrm{d}k_2^2 \frac{\pi}{2\sqrt{4k_1^2 k_2^2 - 2m_{\mathrm{exc}} E_3}} f_{\boldsymbol{k}_1} f_{\boldsymbol{k}_2}$$

$$= \pi |X_{k_3}^{(j)}|^2 |M|^2 m_{\mathrm{exc}} \pi \frac{S^2}{(2\pi)^4} \lambda_{\mathrm{th}}^4 2\pi m_{\mathrm{exc}} k_{\mathrm{B}} T e^{-|E_3|/k_{\mathrm{B}} T}. \tag{6.25}$$

Here $E_3 = E^{(j)}(k_3)$ as usual, and $\lambda_{\mathrm{th}}^2 = 2\pi\hbar^2/(m_{\mathrm{exc}} k_{\mathrm{B}} T)$ is the exciton thermal wavelength.

We have plotted in Fig. 6.5 the linear coefficients a of (6.23) and the quadratic coefficients b of (6.24) as functions of the lattice temperature T, for $k_3 = 0$. The phonon-mediated scattering to the UPs, shown in Fig. 6.5a, first increases rapidly with T, but then increases only linearly for $k_{\mathrm{B}} T > \Omega/2$. The scattering to LPs is weakly dependent on T. Exciton–exciton scattering to UPs and LPs, shown in Fig. 6.5b, is basically the same for the two branches, in agreement with the analytic result of (6.25). The temperature dependence is first exponential, and then weakly decreasing for $k_{\mathrm{B}} T > \Omega/2$. On the same figure we show the analytical results, which compare reasonably well with the numerical ones. The phonon scattering to the LPs, $a_{\mathrm{LP},\boldsymbol{k}=0}$, is strongly sensitive to the q_z cutoff in the phonon scattering, (6.12), and thus to the QW thickness L_z, as shown in the inset of the figure. This is also quite indicative that localization effects are expected to affect this coefficient strongly. The linear and quadratic coefficients show simple exponential dependences on the final energy E, which originate from the Boltzmann factors of the exciton population. We found that they are also well described by the analytic approximations. Again, the phonon scattering to LPs is the exception, and does not show any noticeable structure as a function of E.

The coefficient $b'_{\mathrm{UP},\boldsymbol{k}_3}$ has the same form as $b_{2,\boldsymbol{k}_3}$ of (6.24), where we now consider the final scattering state $\boldsymbol{k}_4 = 0$, and $\boldsymbol{k}_2 = \boldsymbol{k}_3 - \boldsymbol{k}_1$:

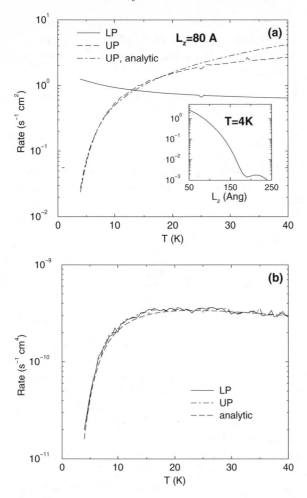

Fig. 6.5. (a) Coefficients for scattering by phonons into LPs and UPs, a of (6.23), and (b) exciton–exciton scattering coefficients into LPs and UPs, b of (6.24), as functions of the lattice temperature T. The analytic approximations introduced in the text are also shown for comparison

$$b'_{\mathrm{UP},\boldsymbol{k}_3} = \frac{\pi}{\hbar} \frac{S}{(2\pi)^2} \int \mathrm{d}\boldsymbol{k}_1 \; |X^{(1)}_{k_1} X^{(1)}_{k_2} X^{(2)}_{k_3} X^{(1)}_{k_4=0}|^2 \; |M|^2$$
$$\times \delta(E_1 + E_2 - E_3 - E_{\mathrm{LP}}) f_{\boldsymbol{k}_1} f_{\boldsymbol{k}_2}$$
$$= \frac{\pi}{\hbar} \frac{S}{(2\pi)^2} |M|^2 |X^{(2)}_{k_3}|^2$$
$$\times \frac{1}{2} \int \frac{dE_1}{\partial_{k^2} E_2} \frac{|X^{(1)}_{k_1} X^{(1)}_{k_2}|^2}{\sqrt{[((k_2 + k_3)^2 - k_1^2)(k_1^2 - (k_2 + k_3)^2)]}} f_{\boldsymbol{k}_1} f_{\boldsymbol{k}_2}.$$
$$(6.26)$$

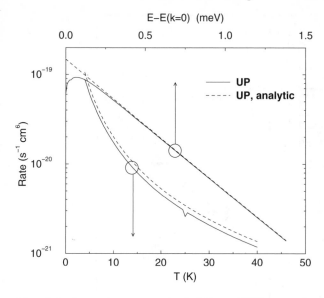

Fig. 6.6. Scattering rate to both LPs and UPs, as a function of the UP energy, and as a function of temperature T, for $T = 4$ K and for $E - E(k = 0) = 0.2$ meV, respectively. The analytic approximations are also shown for comparison

This rate can also be calculated analytically [188], and reads

$$b'_{\mathrm{UP},\mathbf{k}_3} = \pi^2 |X^{(2)}_{\mathbf{k}_3}|^2 \frac{1}{2} \frac{S}{(2\pi)^2} m_{\mathrm{exc}} |M|^2 \lambda^4_{\mathrm{th}} e^{-[E_3 + E^{(1)}(\mathbf{k}=0)]/k_B T}. \qquad (6.27)$$

As $M^2 \propto S^{-2}$, $b' \propto S^{-1}$ and the stimulated scattering is proportional to the LP density. The emission of thermal excitons into both $\mathbf{k} = 0$ UPs and $\mathbf{k} = 0$ LPs is thus proportional to $\lambda^4_{\mathrm{th}}/S$, and becomes a negligible effect at large T and/or large sample surface area. We plot the temperature and energy dependence of this scattering rate in Fig. 6.6. The dependence on the final UP energy E is exponential again, apart from small details.

We now discuss the dynamics which results from (6.20)–(6.22). At small exciton and LP densities, the radiative recombination of excitons is dominant, and the exciton density $n_{\mathrm{exc}} = P_{\mathrm{exc}}\tau_{\mathrm{exc}}$. From (6.21), the UP emission rate has three contributions: a linear contribution to P_{exc} originating from process 2, a quadratic contribution to P_{exc} originating from process 4, and a quadratic contribution to P_{exc} originating from process 5, which is also linear in N_{LP} and thus in P_{LP}. At $T = 4$ K, for $n_{\mathrm{exc}} = 10^9$ cm^{-2}, we obtain a scattering rate of the order of 10^7 s^{-1} for processes 2 and 4. Assuming also $n_{\mathrm{LP}} = N_{\mathrm{LP}}/S = 10^9$ cm^{-2}, we obtain a rate of about 10^8 s^{-1} for process 5, showing that it is measurable and dominates the other scatterings.

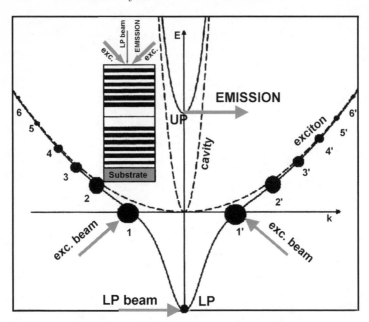

Fig. 6.7. Dispersion of the MEP. Exciton (not to scale) and cavity photon (*dashed lines*) dispersions are shown. The exciton population is depicted by solid circles of proportional size. The numbers are references to typical scatterings, described in the text. *Inset*: the cross section of the microcavity structure. Quarter wavelength dielectric stacks (DBR's) confine the photon inside the cavity, into which a GaAs quantum well, confining the exciton, is embedded. The experimental excitation scheme is also shown

6.6 Stimulated Scattering of Lower Polaritons: Experiment

We have carried out experiments to measure these scattering rates [204, 205]. The setup is shown schematically in Fig. 6.7, where two pump beams from a Ti-sapphire laser ("exciton beams") of intensity I_{exc} at a large angle (55°) and at the exciton energy inject an exciton population, and a third pump beam, from a semiconductor laser ("LP beam"), of intensity I_{LP} at normal incidence and at the LP energy injects LPs. The lasers were continous wave and were focused to a small spot size S of few μm². The microcavity sample was placed in a liquid-He cryostat of nominal temperature $T_{\mathrm{c}} = 4.8$ K. Emission from the UPs was collected at normal incidence and spectrally resolved from the LP emission using a CCD spectrometer.

A typical emission spectrum is shown in Fig. 6.8, for a fixed I_{exc} and varying I_{LP}. The peak of the emission from the UPs is well resolved from the background tails of the emission from the LPs. The spectrum seen around the LP energy is the semiconductor laser spectrum. Even though care was taken geometrically to prevent direct laser light entering the spectrometer,

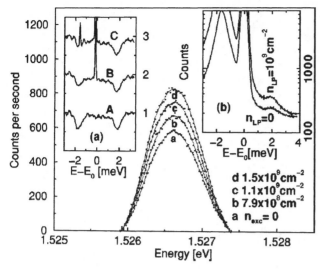

Fig. 6.8. UP emission spectra after the subtraction of the LP and pump tails for different n_{exc} and $n_{\text{LP}} = 1.1 \times 10^9 \, \text{cm}^{-2}$. The *solid lines* are guides for the eyes. *Inset* (a): Reflectivity spectra taken for (A) no pump and probe applied, (B) $n_{\text{exc}} = 2 \times 10^8 \, \text{cm}^{-2}$, no probe applied, and (C) $n_{\text{LP}} = 5 \times 10^9 \, \text{cm}^{-2}$, $n_{\text{exc}} = 2 \times 10^8 \, \text{cm}^{-2}$. Scattered light from the pump and probe can be seen as sharp peaks in the reflectivity spectra, since both the pump and probe are continuous waves. *Inset* (b): The UP emission spectra (raw data), for a fixed $n_{\text{exc}} = 6.5 \times 10^8 \, \text{cm}^{-2}$ and n_{LP} listed on the figure. In *insets* (a) and (b), the energy scale is relative to $E_0 = 1.5247 \, \text{eV}$

some diffusely scattered light from the laser is still detected. The full width at half maximum of the UP emission is constant and close to 0.5 meV. No shifts or additional broadenings were noticed within the experimental range of laser powers, proving that the energy renormalizations are negligible and, in particular, that the Rabi splitting is constant. We came to the same conclusions from reflectivity measurements. The fact that the UP emission shows its typical linewidth rules out any contribution from coherent processes like four-wave mixing, in which case the emission would show a much narrower linewidth related to that of the exciton and LP lasers. This proves that the excitons are thermalized, and that the dynamical picture presented above has significant validity. The UP emission rates were extracted from the raw data by fitting the UP emission peak and subtracting the background tail of the semiconductor laser. Typical results are shown in Fig. 6.9a as a function of I_{exc}, for two different values of I_{LP}. We notice both a linear and a quadratic dependence on I_{exc}, and that the linear dependence is independent of I_{LP}, whereas the quadratic part shows a dependence on it. We fitted the curves in the data set with simple parabolas $dN_{\text{UP}}/dt = c_1(I_{\text{LP}})I_{\text{exc}} + c_2(I_{\text{LP}})I_{\text{exc}}^2$, for each of five different values of I_{LP}. Notice that the offset in Fig. 6.9a has been added for clarity of the figure only. The results for c_1 and c_2 are shown in

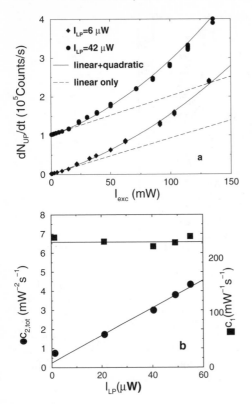

Fig. 6.9. Dependence of UP emission rate on I_{exc} and I_{LP}. (**a**) As a function of I_{exc}, for two different values of I_{LP}, the higher set shifted by 10^5 for clarity. *Dashed lines*: the linear components only. (**b**) the linear and total quadratic dependence on I_{exc}, as a function of I_{LP}. *Solid lines*: fit with a constant for c_1 and with a line for c_2

Fig. 6.9b. c_1 is independent of I_{LP}, whereas c_2 is linearly dependent on I_{LP}. This is clearly in agreement with the dynamical picture presented above, and shows directly evidence of stimulated scattering.

We have tried to quantitatively compare theory and experiment further. To do this, we need to relate the exciton density to I_{exc} and $n_{exc} = \beta I_{exc}$, and the LP number to I_{LP} and $N_{LP} = \gamma I_{LP}$. Direct estimation is difficult, as the mirrors are highly reflective at $55°$ (the measured reflectivity is 99%), the exciton absorption is unknown, and the spot size is not accurately known. Moreover, the exciton lifetime is not accurately known. For these reasons, we carried out an experiment using pulsed laser sources, for both the polariton and the exciton beams [205]. The spectral width of the Ti:sapphire femtosecond laser was narrowed using gratings to about 0.5 meV. The temporal duration of the pulses was about 10 ps. The pulses were temporally aligned. In this way, the exciton and polariton densities could be estimated

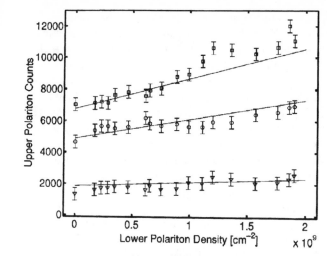

Fig. 6.10. The integrated UP emission intensity as a function of n_{LP}, for varying n_{exc}. The experimental points are indicated by the points with error bars, and the theoretical prediction is shown by the *solid lines*. The pump exciton densities are $1.5 \times 10^9\,\mathrm{cm}^{-2}$ (*squares*), $1.2 \times 10^9\,\mathrm{cm}^{-2}$ (*circles*), and $5.4 \times 10^8\,\mathrm{cm}^{-2}$ (*triangles*)

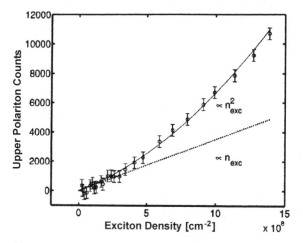

Fig. 6.11. The integrated UP emission intensity as a function of n_{exc}, for a fixed $n_{\mathrm{LP}} = 1.1 \times 10^9\,\mathrm{cm}^{-2}$. Some associated UP emission spectra are shown in Fig. 6.8. The *dotted line* indicates the acoustic phonon scattering contribution to the UP intensity, in the absence of exciton–exciton scattering

independently of their lifetimes, and larger excitation densities were reached. The absolute quantum efficiency of the collection setup was also measured for every run, so as to produce an absolute UP emission count rate. A summary of the results is presented in Figs. 6.10 and 6.11. The lines are theoretical fits, using the calculated scattering rates and compensating for a decaying exciton density. The agreement is reasonable in view of the significant uncertainties in the various measured parameters. It is interesting to notice the significant deviation of the UP emission from linearity in Fig. 6.11. This shows that exciton–exciton scattering becomes larger than exciton–phonon scattering at these densities. Figure 6.10 shows the linearity of the UP emission as a function of the LP density, with a slope that depends on the exciton density, i.e. stimulated exciton–exciton scattering to polaritons. Finally we measured the UP emission as a function of the temporal delay of the polariton pulse with respect to the exciton pulse (Fig. 6.12). From this curve, we subtracted the UP emission that was present without the polariton pulse. Thus, we were left with the stimulated scattering only. We found that this decays at twice the exciton decay rate, giving an independent proof of the binary-scattering origin of this dynamical process.

The continous-wave experiments were carried out at lower excitation densities, and are thus less reliable for making a comparison between theory and experiment, as a larger degree of exciton localization is to be expected. As we discussed before, this type of dynamics was not included in the theoretical analysis, and qualitatively different results are to be expected. The prediction of stimulated scattering is, however, robust, as is the bosonic nature of the lower polariton. The results of these experiments are thus to be interpreted in this way. The quantitative agreement for the results at larger exciton densities obtained using pulsed sources instead of continous-wave sources shows

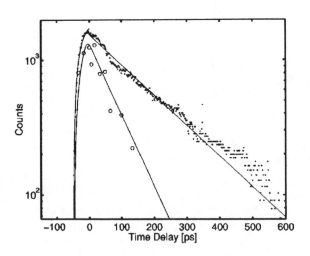

Fig. 6.12. Time dependence of the stimulated scattering and the exciton bottleneck decay, as measured from the LP luminescence. The *solid dots* represent the LP luminescence decay intensity as measured by a streak camera; the *open circles* are the measured stimulated scattering counts into the UP state. The *solid lines* indicate exponential curves with time constants of 96 ps and 190 ps

that, at $n_{\text{exc}} > 10^9$ cm^2, exciton localization is probably less important, and that the free-exciton dynamical picture has significant validity.

6.7 Saturation

For large exciton and LP densities, exciton–exciton scattering to LPs competes with the radiative recombination of excitons. The exciton population is no longer linear with the exciton pump, but becomes depleted by the stimulated scattering process [188]. This behavior is characteristic of a reservoir system depleted by stimulated emission, and its observation would thus indicate that the general dynamical picture presented above is correct. When stimulated scattering of excitons into LPs competes with radiative recombination of excitons, the last three terms of (6.20) become comparable to the radiative-recombination term and cannot be ignored in determining n_{exc}. We can further simplify (6.22)–(6.20) by neglecting the scattering to both UPs and LPs in the depletion dynamics of excitons, for the usual DOS argument suggests that it is a negligible process compared with process 3. The simplified equations read

$$\frac{n_{\text{exc}}}{\tau_{\text{exc}}} = \frac{P_{\text{exc}}}{S} - \left(a_{\text{LP},\text{k}=0}n_{\text{exc}} + b_{\text{LP},\text{k}=0}n_{\text{exc}}^2\right) n_{\text{LP}}, \qquad (6.28)$$

$$\frac{n_{\text{LP}}}{\tau_{\text{LP}}} = \frac{P_{\text{LP}}}{S} + \left(a_{\text{LP},\text{k}=0}n_{\text{exc}} + b_{\text{LP},\text{k}=0}n_{\text{exc}}^2\right) n_{\text{LP}}. \qquad (6.29)$$

The UP emission rate is still given by (6.21). Here we ought to consider scattering to both LPs and UPs, as the large LP population makes this process comparable to all other scattering processes, as explained in the previous section. In Fig. (6.13) we plot the UP emission rate as a function of P_{LP}, for a pump rate $(P_{\text{exc}}/S)\tau_{\text{exc}} = 10^{10}$ cm^{-2} s^{-1}. We have used a thermally averaged exciton lifetime of 100 ps, which is a typical lifetime at $T = 4$ K. At small P_{LP}, the UP emission rate is linear as expected. But at larger P_{LP}, the rate saturates and later starts to decrease. Both the saturation and the subsequent decrease are related to a reduction of n_{exc} with increasing P_{LP}, which we plot in the same figure. Also, we show the corresponding two quantities for a smaller total excitation power $(P_{\text{exc}}/S)\tau_{\text{exc}} = 0.5 \times 10^{10}$ cm^{-2} s^{-1}. Here, the saturation is less dramatic, as expected from the decrease of the exciton–exciton scattering process. This kind of saturation behavior has also been observed in pulsed experiments and reported in [206]. We remark that the saturation described by the rate equations given above does not include phase-space-filling effects and other dynamical effects, which will eventually become important at the highest densities. However, these rate equations give a reasonable description of the saturation effects at lower densities.

We remark again that the results for saturation presented above are qualitative, as, for $n_{\text{exc}} > 10^9$ cm^{-2}, the rate equation approach becomes inaccurate. For $n_{\text{exc}} = 10^{10}$ cm^{-2}, the exciton–exciton scattering rate is about

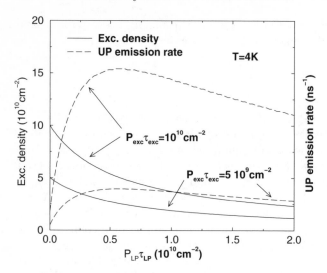

Fig. 6.13. Scattering rate to UPs, and the total exciton density, as functions of the LP density, for $T = 4$ K and two different, fixed exciton pump rates P_{exc}

twice $k_{\mathrm{B}}T$, and the relevant renormalization of the exciton dispersion should set in. The possible onset of superfluidity is also complicated by the issue of localization, which needs further analysis. However, the general qualitative aspects of saturation that we have presented here have a very simple physical basis and should persist even in a more accurate model of the dynamics. We also notice that the condition $\Gamma_{\mathrm{exc-exc}} < \Omega_{\mathrm{R}}$ is much less restrictive, and it has been experimentally estimated that the Rabi splitting collapses at $n_{\mathrm{exc}} \sim 4 \times 10^{10}$ cm^{-2} [202]. Thus, the neglect of Rabi-splitting renormalization in the range of densities presented above is still well justified, because the collapse of the splitting is not gradual, but abrupt [203].

7. Stimulated Emission of Exciton Polaritons

7.1 Introduction

We can draw important conclusions from the results of the previous chapter, in which the stimulated scattering of reservoir excitons into polaritons was presented. First, the exciton polaritons show an evident bosonic nature. Second, the exciton–exciton scattering is well characterized; its strength has been measured and is in qualitative agreement with theory. Third, the influence of the interface disorder of the quantum well surfaces on the diffusion of the lower polaritons is very limited. The lower polaritons are thus good candidates for the realization of a bosonic laser, or boser. This system, although bearing strong underlying similarities to the laser, has fundamental interest. In fact, the bosonic particles undergoing the stimulated emission process are not simple photons, but contain a large excitonic component, which is a material component. Thus, a rare example of stimulated emission of matter waves could be realized, in addition to that obtained by the condensation of bosonic atoms in ultracold traps.

In a system where polariton loss is made negligible compared with the thermalization rate, the polaritons become thermalized. The exciton polariton system in the microcavity is two-dimensional. At low density it is weakly interacting, and in a first approximation we may neglect the resulting perturbation of the energy levels. The population distribution follows Bose–Einstein statistics, and the population N_0 of the lowest-energy state, which has $k = 0$, is given by

$$N_0 = n_{\text{pol}} \lambda_{\text{th}}^2 = n_{\text{pol}} \frac{2\pi\hbar^2}{m_{\text{pol}} k_B T}, \tag{7.1}$$

where n_{pol} is the exciton polariton density and λ_{th} is their thermal wavelength. As the exciton polariton mass is at least three orders of magnitude smaller than the exciton mass, a large advantage in terms of the condensation temperature and/or condensation density is obtainable. From this perspective, the realization of stimulated emission relies on the achievement of a sufficient density and a low temperature. In the present argument, the thermalization hypothesis is not addressed, but it is clearly a central point, on which the whole argument depends. It is therefore necessary to address the more general problem of finite loss of the lowest polariton states. In this case the system becomes an open system. When the cooling of hot excitations into lossy lowest-energy states is viewed as a gain process, an analogy to the conventional laser system is established. There, the lossy photon mode of a cavity is fed by the photons emitted from a reservoir of inverted atoms. Eventually, the resulting gain compensates for the loss, and stimulated emission results.

The loss of the microcavity polaritons mainly originates from their escape from the cavity through the enclosing mirrors. It is thus related to the cavity photon component of the polariton and to the cavity quality factor Q. The loss rate of the current cavities is very large, in the range of ps^{-1}, and easily dominates any other loss process. The resulting small Q factors of a few thousand are only a technological limitation, related to the quality of the dielectric-mirror interfaces produced by the epitaxial growth process. A different technology for deposition of the dielectric mirrors (sputtering) results in cavities having quality factors well over three orders of magnitude larger (super-Q cavities).

Concerning the gain for the microcavity polaritons, and the analog of the inverted-atom reservoir of the conventional laser, the simulations have shown that a thermal reservoir of excitons is produced when excitons are injected into the microcavity. We remark again that experimental results relating to the stimulated scattering have also shown indirectly the existence of this reservoir. Recent experiments with calibrated angle-resolved photoluminescence have also directly shown the existence of this reservoir. Gain is provided by cooling of this exciton population into the lower polaritons. We have identified two different types of relaxation mechanism, phonon emission events and exciton–exciton scattering events. We have shown that at large densities the exciton–exciton scattering process dominates the phonon process of relaxation into lower polaritons. Even if gain increases with exciton density, however, there is an upper limit on it, and beyond a certain loss, no stimulated emission can be produced. When $n_{exc}a_B^2$ is of the order of unity, excitons ionize spontaneously, and the plasma phase is produced (a Mott transition). At smaller exciton densities, the strong exciton–photon interaction is washed out by collisions in the exciton gas, and a transition from strong to weak-coupling occurs. This transition is usually observable in the optical spectrum as a collapse of the Rabi splitting. In GaAs-based structures it occurs around $n_{exc} \sim 4 \times 10^{10}$ cm^{-2}, as shown in [202]. This is the same density at which the exciton–exciton scattering rate equals the Rabi splitting. We thus find that at this critical density, the gain from phonon emission amounts to a small fraction of 1 ps^{-1} – typically a few ns^{-1} – , and is very dependent on the quantum well width. The gain from exciton–exciton scattering amounts to just a fraction of 1 ps^{-1} – typically a few times (10 ps)$^{-1}$ – depending on the bath temperature, and is thus insufficient to compensate for the typical losses of the order of 1 ps^{-1}. The resulting polariton population thus remains well below unity when the Rabi splitting collapses.

We have performed a simulation of the stationary exciton polariton distribution in a typical GaAs microcavity with 4 meV of Rabi splitting, zero detuning between the excitons and the cavity, and a cavity photon lifetime of 0.6 ps. Results for various densities are shown in Fig. 6.4. Even at the largest density considered, $n_{\mathrm{exc}} = 5 \times 10^9$ cm^2, the lowest-energy polaritons have a population well below unity. Most of the strongly coupled modes that have negative energy (measured with respect to the bottom of the exciton band) have in fact a population that is significantly below the value given by the Boltzmann distribution (a straight line at large energy on a logarithmic plot). It is interesting to observe that some polariton modes that have $k \neq 0$ reach a population larger than one (degenerate statistics). When analyzing this result, we found that it was due to their longer lifetime with respect to the $k = 0$ polariton, as these modes have a decreased cavity photon content. This result indicates that even a slightly longer polariton lifetime would be sufficient for observing degenerate statistics. As remarked above, the cooling in this region of density is to be attributed to exciton–exciton scattering.

In the following sections of this chapter we thoroughly analyze the stimulated emission of lower polaritons in appropriate conditions. In order to make a closer analogy to the conventional laser, we consider a structure where one or a few of the lower polaritons are selected by confinement in an in-plane cavity. The conceptual realization of this in-plane cavity – by etching a planar microcavity into a small post structure – is rather straightforward. In such a system, it is also possible to study the statistical properties of the stimulated emission of polaritons with the well-established techniques developed for conventional lasers. Establishing the analogies with and differences from this latter prototype system constitutes a significant advancement in the understanding of the physics of this boser system. As a secondary advantage, we can also expect an increased exciton content of the lowest confined polariton mode, and therefore a longer lifetime. This is an important practical advantage for the achievement of stimulated emission.

7.2 Rate Equations and Threshold Behavior

We consider a post of square section with side D, as shown in Fig. 7.1. Photons are well confined in the post by the large difference of index of refraction between the air and the semiconductor. Thus, in a lowest-order approximation, we can totally neglect the small effect of the photon wavefunction extending beyond the semiconductor material into the air, and assume that the in-plane wavevectors k_x and k_y are quantized to $n_x \pi / D$, and $n_y \pi / D$, respectively, with $n_x = 1, 2, \ldots$ Then, the energies of the micropost polariton modes are those of the planar-microcavity polaritons at the quantized in-plane momenta k_x, k_y. A schematic representation of the micropost polariton modes is shown in Fig. 7.2.

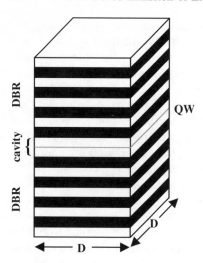

Fig. 7.1. The post structure. The distributed Bragg reflectors, the cavity and the embedded quantum well are indicated

Within the same assumption of good lateral confinement, the radiative recombination rates are given also by those of the planar cavity at the quantized in-plane wavevectors. In particular, the lowest confined polariton has $k_x = k_y = \pi/D$, and a much longer lifetime than the $k = 0$ mode, owing to an increased exciton content and reduced cavity photon content. In the structure considered here, having $D = 3\,\mu\text{m}$, with a cavity photon lifetime of 0.6 ps, this lifetime is about 20 ps, and it can be even longer for smaller posts. However, we want somehow to take into account finite leakage of the polaritons through the sidewalls, owing, for example, to surface roughness, and other nonradiative recombination processes for the exciton. For this purpose, we have used a shorter effective polariton lifetime of 10 ps in the calculations. As we can see from Fig. 7.2, the exciton modes are closely spaced in energy and practically not affected by confinement in the large cavity, as their mass is three orders of magnitude larger than the photon mass. Thus, the radiative lifetime of the excitons is also unperturbed by the in-plane confinement in the post structure. Of course, some nonradiative recombination centers can be introduced by the etching process, but we do not address technological issues here. The radiative recombination of excitons in a microcavity occurs through the leaky modes of the dielectric mirrors, and has been calculated in detail in [179].

We describe the dynamics of the populations with the same rate equations as in the previous chapter, (6.3), modified to take into account the lateral confinement of the lowest confined polariton modes. In particular, for the scattering to and from a confined polariton mode, we can use the expression for the scattering matrix elements (6.17). The density of states of the confined mode is $\partial E(k)/\partial k^2 = \Delta E S/(4\pi)$, where $S = D^2$ is the surface area of

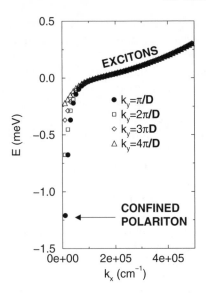

Fig. 7.2. The quantized energy/momenta for a 3 μm × 3 μm post structure. There is a lowest confined polariton mode, with an energy $-|E_0|$, and densely packed modes at positive energy, which constitute the exciton reservoir

the post. This substitution is exact for the lowest confined mode, for which $k_x = k_y = \pi/D$; it accounts for only an azimuthal average of the scattering rates for the higher confined polariton modes, but these are not considered further in this work.

From the results of the previous chapter, we know that phonon scattering from the exciton reservoir to the polaritons is very weak, as it is suppressed by the reduced density of states of the polaritons. Instead exciton–phonon scattering within the exciton reservoir is important in driving the exciton temperature to the lattice temperature. The exciton–exciton scattering rate is dependent on the exciton density, and becomes significant at large densities. We have characterized the strength of the exciton–exciton scattering by the out-scattering rate from the $k = 0$ exciton $\hbar\Gamma_{exc-exc} \sim (\pi^2 E_B^2/2E_L)(n_{exc}a_B^2) \sim 1$ meV when $n_{exc} = 10^{10}$ cm^{-2}. Here $E_L = k^2/(2m_{exc}a_B^2)$ is a typical exciton kinetic energy. As has been explicitly shown in the previous chapter, the relaxation of reservoir excitons into polaritons through exciton–exciton scattering is not suppressed by a reduction of the density of states as in the case of the phonon emission process. We recall that in this binary relaxation, two excitons scatter into a polariton and a hotter exciton. The final density of states is thus the large density of states of the excitons. We gave a detailed estimate of this scattering rate in (6.25), considering the two initial excitons to be in a thermal distribution and the final polariton to be at an energy $-\hbar\Omega$ and momentum zero. We recall here that an exponential Boltzmann factor $\exp(-\hbar\Omega/k_B T)$ appears in the expression for this rate. The heuristic motivation is simple, and is to be found in the energy and momentum conservation laws which hold for the binary scattering. In fact, an energy $\hbar\Omega$ is released to the excitons in the scattering process. If one of the initial excitons has a small energy, and thus a small momentum, a small total momentum is exchanged in the scattering. The other initial exciton then needs to have a very large momentum, so that its small momentum change in the scattering involves a large change of energy $\hbar\Omega$. But this is unfavorable, as few high-energy excitons are found, according to the Boltzmann distribution. The most favorable situation starts with both excitons having an energy $\hbar\Omega/2$, so that the above Boltzmann factor results. In conclusion, we understand that if $k_B T \sim \hbar\Omega$, the relaxation from exciton–exciton scattering into a polariton can be significantly fast, of the order of a fraction of 1 ps^{-1}, at the critical exciton density, as it can be deduced from the values shown in Fig. 6.5. Thus, significant losses can be compensated, and stimulated emission of the polaritons eventually observed. We have, finally, to remark that when $k_B T \sim \hbar\Omega$, the reverse process of polariton–exciton scattering to final exciton states also becomes significant. However, polariton-to-polariton scattering is suppressed by the density of states argument, as has also been shown experimentally in [207]. The reverse scattering process is a heating process and is thus weaker than the relaxation process (cooling), as established by detailed-balance arguments. In fact, when both the in-scattering and the out-scattering rates are significantly larger than the other losses, thermalization of the polariton system with the exciton reservoir will be attained.

In Fig. 7.3 we present the numerical results of the integration of the rate equation (6.3) for the micropost system, under stationary conditions $d/dt = 0$. As in the previous chapter, we have assumed direct injection of excitons by an optical pump. The excitons thermalize to a temperature T_{exc} which is close to the lattice temperature T, below threshold. At large pumping rates, we observe a clear threshold behavior in the lowest-polariton population close to $N_0 = 1$, and moderate clamping of the exciton density. In Fig. 7.3b, we observe that this threshold is related to sizeable heating of the excitons, which is most apparent for the lowest temperature considered, $T = 2$ K. This temperature increase shows that exciton–exciton scattering is the origin of the gain. In fact, we remarked above that an energy of about $|E_0| = 1$ meV is released to the excitons for every polariton created by an exciton–exciton scattering process. This excess energy is dissipated through phonon emission events within the exciton reservoir. The stationary exciton temperature results from the energy inflow and outflow balance. The energy inflow rate is thus 1 meV $\times N_0/\tau_0 = 0.1 \times N_0$ meV ps^{-1}. The phonon absorption and emission rates are balanced when $T_{\text{exc}} = T$. Then, to lowest order, the net energy outflow rate for the phonon scattering process is proportional to $n_{\text{exc}}(T_{\text{exc}} - T)$. As n_{exc} is roughly clamped above threshold, we finally obtain $(T_{\text{exc}} - T) \propto N_0$. We conclude that exciton heating is an essential fingerprint of gain from exciton–exciton scattering, and its observation would

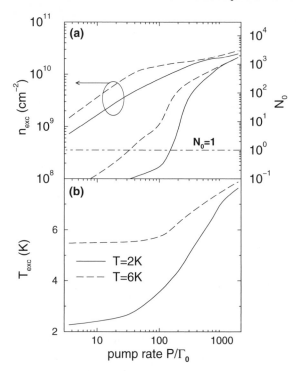

Fig. 7.3. (a) Exciton density, *left*, and polariton population, *right*, and (b) exciton temperature versus pump rate for two different temperatures, for $\tau_0 = 10$ ps and a 3 μm ×3 μm post structure

provide a strong evidence of this origin for the gain. Finally, we comment on the origin of the partial clamping of the exciton population above threshold. As we observed above, an increase in the exciton temperature corresponds to an increase of the polariton heating by exciton collisions (the inverse of the cooling process). Therefore, an increase in the exciton temperature has to be compensated by an increase of the exciton density to achieve threshold.

In order to substantiate the analysis of the temperature dependence of the relaxation rate of excitons into polaritons through binary collision events, we have calculated the exciton density at threshold as a function of the lattice temperature. We define the threshold by the condition $N_0 = 1$. We show the results in Fig. 7.4. It is clear that an optimal temperature exists, which is of the order of $|E_0|$. Moreover, the density at threshold increases monotonically for larger temperatures, as expected from the increase of the inverse process of exciton-into-polariton relaxation. We have also studied the dependence of the exciton density at threshold as a function of the lifetime of the confined polariton. We plot the results in Fig. 7.5. We find a monotonic decrease of the threshold density with the confined-polariton lifetime, as expected. We remark that the threshold density clearly shows an asymptotic limit.

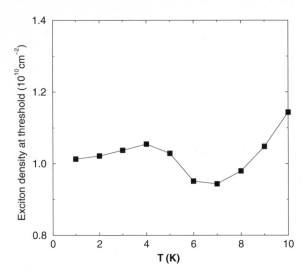

Fig. 7.4. Exciton density at threshold versus lattice temperature. Other parameters as in Fig. 7.3

This is the limit of the threshold in thermal equilibrium, which of course occurs at a finite density too. This limit is easily analyzed, by considering a single confined level at energy $-|E_0|$ below the exciton reservoir. Using a Bose–Einstein distribution for the population in the exciton reservoir and the confined polariton, we find that $N_0 = 1$ occurs when

$$n_{\mathrm{exc}}\lambda_{\mathrm{th}}^2 = \frac{1}{2}\mathrm{e}^{-|E_0|/k_B T}. \tag{7.2}$$

Here λ_{th} is the thermal wavelength of the *exciton* reservoir. Using $k_B T \sim |E_0|$ and the typical exciton mass in GaAs quantum wells, $m = 0.25 m_0$, we find typical densities at threshold of the order of 4×10^9 cm^{-2} or larger. When $k_B T \ll |E_0|$, within the assumption of thermal equilibrium, the exciton gas becomes very dilute, and most of the population is found in the lower-polariton mode. However, we have noticed that in this condition, the efficiency of exciton–exciton scattering as a relaxation process significantly decreases, both because of the reduced temperature and because of the reduced exciton density. Thus, the achievement of threshold in the lower-polariton population in this very-low-density condition requires a much longer lifetime of these modes. In particular, the slow phonon relaxation process would eventually dominate the exciton–exciton scattering process in this condition, and would alone determine the exciton density at threshold. In the planar case, an analogous situation is expected. In particular, lower-polariton condensation driven by exciton–exciton scattering is likely to occur only with a significant exciton density, above 4×10^9 cm^{-2}, unless the polariton lifetimes are very long. Here also, eventually the phonon-mediated relaxation dominates. This

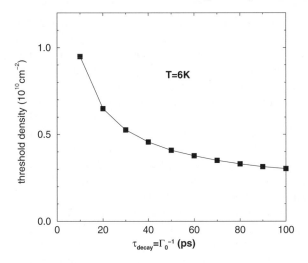

Fig. 7.5. Exciton density at threshold versus the total polariton lifetime. Other parameters as in Fig. 7.3

may not be practically observable until polariton lifetimes of the order of nanoseconds in GaAs structures are realized.

7.3 Langevin Equations and Noise Properties

In this section we shall calculate the statistical properties of the lowest polariton state under stationary conditions. We expect that the general characteristics found in the prototype laser system are also found here, in particular, the observation of Poissonian number statistics far above threshold. The occurence of this type of statistics is related to the existence of a reservoir of finite size, when feedback effects on the gain become relevant and suppress the large excess noise produced in the stimulated emission process. The rate equations used before for characterizing the threshold and the population dynamics are clearly insufficient for this purpose, as only statistical averages are explicitly addressed there. We need to take into account the fluctuations around the average values. We must resort to an effective, perturbative method, extending the rate equations considered before. The Langevin equations provide a convenient approach, which works accurately both well below and well above threshold. The polariton and exciton states are then treated as quantum mechanical operators, and their evolution is calculated perturbatively up to second order in the interaction Hamiltonian. In order to recover the correct commutation relations of the bosonic operators, appropriate white-noise sources are added to the resulting equations [208]. In principle, we need an equation for each exciton state. However, we know that these states form a reservoir. Physically, this means that scattering processes within the reservoir are faster than any other scattering process with the external world. In particular, this concerns scattering with the lasing, lower-polariton mode. Eventually, at some point this reservoir assumption breaks down when $N_0 \gg 1$, as scattering to the lasing mode becomes stimulated and faster than the scattering processes within the reservoir itself. We know that in a bosonic system in thermal equilibrium, fast scattering between the condensed ground state and the finite-energy states leads to the Bogolyubov phonon-like quasiparticles [209]. In this case, the nature of the reservoir states clearly changes. The fundamental parameter characterizing the reservoir–ground state interaction is $\mathcal{W}N_0$, where $\mathcal{W} = 6E_\mathrm{B}a_\mathrm{B}^2/S$ is the exciton–exciton interaction. The other energy scale charactering the reservoir and ground state is $|E_0|$. When $\mathcal{W}N_0/2 < |E_0|$, the renormalization of the exciton reservoir is negligible. For $D = 3\ \mu\mathrm{m}$ we obtain $N_0 < 3 \times 10^3$. The fast scattering processes within the reservoir also establish a thermal equilibrium, which we have seen to be effectively reached in the numerical simulation using the rate equations. We are therefore allowed to effectively neglect the deviation of the exciton population distribution from the thermal distribution, and we can use the total exciton number as the unique operator characterizing the reservoir. Moreover, to a first approximation we can assume that the exciton temperature is fixed, and we neglect the heating effects related to exciton–exciton scattering. The importance of the temperature fluctuations needs to be established, whereas the average effects calculated in the previous section can be easily included. The Langevin equations finally obtained read [210]

$$\dot{b}_0 = \frac{G - \Gamma_0}{2} b_0 - \mathrm{i}\frac{\mathcal{W}}{2}(|X_0|^2 N + |X_0|^4 N_0)b_0 + F_{b_0} + F_{\Gamma_0}, \tag{7.3}$$

$$\dot{N} = p - \Gamma N - GN_0 - R_\mathrm{sp} + F_\Gamma + F_p + F_N. \tag{7.4}$$

Here b_0 is the destruction operator for the polariton mode, $N = \sum_{\boldsymbol{k}} n_{\boldsymbol{k}}$ is the total number of excitons, p is the pumping rate, G is the net gain, i.e. the net in- and out-scattering from the polariton mode, and $|X_0|^2$ is the exciton content of the polariton mode. The spontaneous-emission terms R_sp are just the in-scattering terms into the polariton mode. Γ is the average of the exciton recombination rate over the thermal distribution. The first term on the right-hand side of (7.3) is the net gain minus the loss of the laser oscillator. The last two terms are related to energy shifts of the polariton level and do not appear in the rate equations. The first term $|X_0|^2\mathcal{W}N$ can be viewed as a modulation of the index of refraction of a conventional laser cavity by a change of carrier density. The other term $|X_0|^4\mathcal{W}N_0$ is directly derived from the microscopic Hamiltonian interaction term $\hbar\mathcal{W}|X_0|^4 b_0^\dagger b_0^\dagger b_0 b_0/4$. As written here, it is analogous to the energy modulation provided by a Kerr medium. This is unique to this polariton laser system, as such a term does not usually exist when b_0 represents a photon in the vacuum. The $F(t)$ are white-noise sources and have the following statistical properties:

$$\langle F_{b_0}^{\dagger}(t)F_{b_0}(t')\rangle = \bar{R}_{\mathrm{sp}}\delta(t-t'), \tag{7.5}$$

$$\langle F_{b_0}(t)F_{b_0}^{\dagger}(t')\rangle = (\bar{R}_{\mathrm{sp}} - \bar{G})\delta(t-t'), \tag{7.6}$$

$$\langle F_{\Gamma_0}^{\dagger}(t)F_{\Gamma_0}(t')\rangle = 0, \quad \langle F_{\Gamma_0}(t)F_{\Gamma_0}^{\dagger}(t')\rangle = \Gamma_0\delta(t-t'), \tag{7.7}$$

$$\langle F_p(t)F_p(t')\rangle = p\,\delta(t-t'), \tag{7.8}$$

$$\langle F_N(t)F_N(t')\rangle = [(\bar{R}_{\mathrm{sp}} - G)\bar{N}_0 + \bar{R}_{\mathrm{sp}}(1+\bar{N}_0)]\delta(t-t'), \tag{7.9}$$

$$\langle F_\Gamma(t)F_\Gamma(t')\rangle = \Gamma\,\bar{N}\,\delta(t-t'). \tag{7.10}$$

The overbars represent statistical averages. Cross correlations are found only between the forces originating from the scattering between the polaritons and the exciton reservoir. Introducing $N_0 = b_0^{\dagger}b_0$, we have

$$\langle F_{N_0}F_{N_0}\rangle = \langle F_N F_N\rangle = -\langle F_N F_{N_0}\rangle. \tag{7.11}$$

This equation is exact, as scattering conserves the total number of particles. The Langevin equations (7.3) are very similar to those of a conventional semiconductor laser [211], except for the different gain mechanism, and the last self-energy term in (7.3) as explained above. Solution of these equations is usually carried out for $\bar{N}_0 \gg 1$, with a linearization procedure of the operators around their average values [211]. In this case, phase fluctuations are considerably slower than amplitude fluctuations. In this limit, it is possible to introduce the Hermitian amplitude and phase operators δb_0 and ϕ, respectively, by means of

$$b_0 \equiv (\bar{b}_0 + \delta b_0)\mathrm{e}^{-\mathrm{i}\phi},$$

with $\bar{b}_0 = (\langle b_0 + b_0^{\dagger}\rangle)/2$. We also introduce $N = \bar{N} + \delta N$ and $\delta N_0 = 2\bar{b}_0\delta b_0$. We obtain

$$\dot{\phi} = \frac{\mathrm{i}}{2\bar{b}_0}\left(F_{b_0} - F_{b_0^{\dagger}} + F_{\Gamma_0} - F_{\Gamma_0}^{\dagger}\right)$$
$$+\frac{3|X_0|^4}{4}W\,\delta N_0 + \frac{|X_0|^2}{2}W\,\delta N + \frac{|X_0|^2}{2}W\bar{N}\frac{\delta N_0}{\bar{N}_0}, \tag{7.12}$$

$$\dot{\delta N_0} = \bar{b}_0\left(F_{b_0} + F_{b_0^{\dagger}} + F_{\Gamma_0} + F_{\Gamma_0}^{\dagger}\right) + \eta\delta N, \tag{7.13}$$

$$\dot{\delta N} = F_\Gamma + F_p + F_N - (\Gamma+\eta)\delta N - \Gamma_0\delta N_0. \tag{7.14}$$

Here $\eta = \bar{N}_0\,\mathrm{d}\bar{G}/\mathrm{d}\bar{N}$. Let us also remark that we have neglected the average energy level shifts which appear in the phase equation (7.12), and which typically amount to fractions of a meV at most in the cases considered. This term represents the change of the lasing frequency with carrier density. As is customary in laser theory, and in agreement with the previous observation that far above threshold the phase is slowly fluctuating and decouples from the faster number fluctuations, we have also neglected the phase operators in the number equations. The new self-energy term, $(3/4)|X_0|^4 \mathcal{W} \delta N_0$, is not contained in the usual laser theory, as remarked before. It represents a self-phase modulation, in contrast to the amplitude-to-phase coupling (also found in conventional lasers), which is always mediated by modulation of the carrier density, and corresponds here to the other term, $|X_0|^2 \mathcal{W} \delta N/2$. Equations (7.13) and (7.14) were solved by standard Fourier transform techniques, and the power spectra were calculated, taking into account the autocorrelation and crosscorrelation of the noise sources. The laser was defined to be very far above threshold, so that the confined-polariton population exceeds the exciton population in the reservoir, $\bar{N}_0 \gg \bar{N}$. Neglecting for simplicity the reverse process of exciton–exciton to polariton–exciton scattering (polariton heating), we have $\bar{G} \sim \Gamma_0 \sim \Gamma'_{e-e} \eta^2 \bar{N}^2$ and $\eta \sim 2\bar{G}\bar{N}_0/\bar{N} \gg \Gamma_0, \Gamma$. The power spectrum of $\delta \tilde{N}_0(\omega)$ can be found by Fourier transformation, and in this limit it is simply Lorentzian, of width Γ_0, and the total noise power $\langle (\delta N_0)^2 \rangle \sim \bar{N}_0$. In this limit, the usual Poissonian fluctuations of the lasing confined polariton are recovered.

We have to remark at this point that, with the parameters used in the present numerical examples, $D = 3 \, \mu m$ and a polariton lifetime of 10 ps, the exciton density at clamping is around 2×10^{10} cm^{-2}. When $\bar{N}_0 \sim \bar{N}$, a total excitation density close to 4×10^{10} cm^{-2} is obtained, where the Rabi splitting collapses. Therefore, for GaAs-based cavities with these parameters, it is not possible to reach the far-above-threshold condition. Longer lifetimes are needed for this purpose. However, we have calculated a significant reduction of the super-Poissonian noise of stimulated emission around threshold, $\langle (\delta N_0)^2 \rangle \sim \bar{N}_0(1 + \bar{N}_0)$, at smaller densities and populations.

7.3.1 Polariton Squeezing

Besides the usual phase diffusion resulting from the spontaneous emission, which gives rise to the Schawlow–Townes linewidth of the laser, we have contributions from the usual phase modulations resulting from fluctuations of the exciton population, and from the self-modulation. We noticed above that, far above threshold, the fluctuations of N_0 dominate over those of N. In this case, the self-phase modulation dominates over the other phase diffusion mechanisms discussed earlier. Then, the frequency fluctuations are proportional to the number fluctuations, and information about the former may be used to reduce the latter. In particular, we consider a linear combination of the amplitude and frequency fields, given by the new operator

$$S = \delta b_0 + c \frac{\bar{b}_0 \dot{\phi}}{\Gamma_0} = \frac{\delta N_0}{2\bar{b}_0} + c \frac{\bar{b}_0 \dot{\phi}}{\Gamma_0}, \tag{7.15}$$

where $c \ll 1$ is an arbitrary real number. Then, the power spectrum of S is calculated and the number c chosen to minimize its noise power. We find that by choosing $c \sim -3\mathcal{W}/(2\Gamma_0 \bar{N}_0 |X|^4)$, perfect squeezing is achieved. This of course neglects the background phase diffusion processes, which, however, give only a small contribution proportional to c^2. The detection of an internal field, and therefore of the number S, is not straightforward, as we cannot place detectors inside the cavity. The result directly shows, however, that the frequency carries all the information on the polariton number fluctuations. This information is easily retrieved and used through a linear mixing process, which generates a noiseless state of polaritons in the microcavity.

Alternatively to internal detection, we may consider the statistical properties of polaritons extracted from the microcavity. For example, polaritons could tunnel to an external exciton reservoir. In all cases, they eventually leave the microcavity as external photons. We have analyzed only this possibility. We obtain a minimum noise power of $1/4$, using a value of c that is half of that for optimal internal squeezing. This is a moderate 3 dB squeezing with respect to the standard quantum limit of $1/2$. This result has a simple physical interpretation. Polariton number fluctuations can result equally well from the radiative loss process and from pump fluctuations, which we considered to be Poissonian (incoherent pumping) in (7.8). The fluctuations originating from the radiative loss process cancel outside the cavity, within the cavity bandwidth, by beating with the external noise force F_e [211]. Usually, we are thus left with pump noise only, within the cavity bandwith. Clearly, the phase fluctuations far above threshold in the micropost system carry the information about the polariton number fluctuations. If, through the usual amplitude–frequency mixing process, we cancel completely the number noise fluctuation, we are left with the standard noise from F_e. The best we can do is therefore to use half the value of c for optimal internal squeezing: half of the fluctuations from the radiative decay will be canceled in the beating with F_e, and half of the pump fluctuations will still be left. This finally makes $1/4$ of the noise from the radiative recombination process and $1/4$ of the pump noise, adding up to a total noise power of $1/4$.

As remarked before, in the GaAs structures considered in this work, we do not expect that the far-above-threshold condition can be fully reached. We have evaluated numerically both the internal and the external optimal noise powers which can be obtained with posts of various sizes. A moderate degree of squeezing could be obtained for the internal field only. However, its origin is not in the self-phase modulation. The noise suppression close to threshold mainly originates from the standard phase modulation, whereas at larger N_0, it mainly originates from the self-phase modulation but is degraded by the standard phase modulation. This is rather obvious, as the fluctuations in N_0 are small close to threshold, and do not provide significant phase modulation compared with that originating from the reservoir number fluctuations. In conclusion, the observation of polariton squeezing requires one to reach the far-above-threshold condition, and this will presumably be possible only in different materials or structures with longer polariton lfetimes.

7.3.2 Emission Linewidth

Another interesting property of the conventional laser is the reduction of the linewidth of the emission as $(\bar{N}_0)^{-1}$. This is the typical Schawlow–Townes behavior, and is simply related to the purely imaginary noise source of (7.12) and the clamping of the spontaneous emission rate $\bar{R}_{\rm sp}$ above threshold. Amplitude-to-phase coupling produces an enhancement of the phase fluctuations and thus of the laser linewidth. The spectral shape of the emission can be calculated with the following correlation function [208]:

$$I(\omega) = \int_0^\infty d\tau \, e^{-i\omega\tau} \langle A^\dagger(\tau)A(0)\rangle. \tag{7.16}$$

Here $A(\tau)$ is the electromagnetic-field amplitude at the detector position, outside the cavity. Vacuum field fluctuations clearly do not contribute to the average, and we may therefore substitute $b_0(\tau)$ for $A(\tau)$ in the above, apart from the trivial delays produced by propagation of light from the cavity to the detector.

Below or around threshold, the system behaves as a simple harmonic oscillator, and we may neglect the amplitude-to-phase couplings to a first approximation. The average in (7.16) is then given by the quantum regression theorem [208], and we obtain a Lorentzian lineshape for $I(\omega)$, of width $\Gamma_0 - \bar{G} = \bar{R}_{\rm sp}/\bar{N}_0$. Below threshold, this linewidth is simply Γ_0. Above threshold $\bar{R}_{\rm sp}$ is approximately clamped, and an $(\bar{N}_0)^{-1}$ linewidth reduction is found. In order to calculate the contributions of the amplitude-to-phase couplings to the linewidth above threshold, we have to resort to the linearization procedure introduced above, which holds for $\bar{N}_0 \gg 1$. The lineshape can then be connected with the phase diffusion coefficient, and

$$I(\omega) \propto \int_0^\infty d\tau \, e^{-i\omega\tau} e^{-\langle[\phi(\tau)-\phi(0)]^2\rangle}. \tag{7.17}$$

In calculating $\langle[\phi(\tau) - \phi(0)]^2\rangle$, we find that the imaginary noise sources in (7.12) decouple from the number fluctuations, and separate contributions result. The first term produces again the Schawlow–Townes linewidth, as calculated before. The calculation of the contribution of the number fluctuations to the linewidth is more involved, as these operators have colored spectra, and are correlated. Using

$$\langle[\phi(\tau) - \phi(0)]^2\rangle = \int_0^\tau dt_1 dt_2 \langle \dot{\phi}(t_1)\dot{\phi}(t_2)\rangle,$$

inserting (7.12) in this equation, and passing to Fourier space, we may use the resulting solutions for $\delta N(\omega)$ and $\delta N_0(\omega)$. We can analyze the contributions from the fluctuations of N and N_0 separately, and consider the far-above-threshold condition, $\bar{N}_0 > \bar{N}$, where the operators have a simple Lorentzian power spectrum. The last term in (7.12) gives negligible contributions in this regime. From semiconductor laser theory, we know that $|X_0|^2 \mathcal{W} \delta N/2$ gives an enhancement factor α^2 to the Schawlow–Townes linewidth. This can be calculated to be

$$\alpha^2 = \frac{\Delta\omega_{\text{enh}}}{\Delta\omega_{\text{ST}}} \sim \frac{|X_0|^2 \mathcal{W}^2 \bar{N}^2}{4\Gamma_0^2}. \tag{7.18}$$

For example, using the parameters of the numerical example given before, an enhancement factor $\alpha^2 \sim 10$ is found. This value is not uncommon in semiconductor lasers.

The contribution of the $|X_0|^4 \mathcal{W} \delta N_0$ term to the laser linewidth is qualitatively new. Using the Lorentzian power spectrum of δN_0 far above threshold, we find

$$\langle [\phi(\tau) - \phi(0)]^2 \rangle_{\text{SE}} = \frac{9|X_0|^4 \mathcal{W}^2 4\bar{N}_0}{\Gamma_0}\left(\tau - \frac{1 - e^{-\Gamma_0\tau}}{\Gamma_0}\right). \tag{7.19}$$

This correlation produces a nonexponential decay of the integrand of (7.17), and two regimes can be identified. A simple exponential decay dominates when $9|X_0|^4 \mathcal{W}^2 \bar{N}_0 / 4\Gamma_0^2 \ll 1$; otherwise, the decay of the initial coherence is of Gaussian type. The resulting emission lineshapes are Lorentzian and Gaussian, respectively, of widths

$$\Delta\omega_{SE} = \begin{cases} \dfrac{9|X_0|^4 \mathcal{W}^2 \bar{N}_0}{4\Gamma_0}, & 3|X_0|^2 \mathcal{W} \sqrt{\bar{N}_0} \ll 4\Gamma_0 \\ \dfrac{3\ln 2|X_0|^2 \mathcal{W} \sqrt{\bar{N}_0}}{2}. & 3|X_0|^2 \mathcal{W} \sqrt{\bar{N}_0} \gg 4\Gamma_0 \end{cases} \tag{7.20}$$

What is most interesting is that both linewidths *increase* with $\bar{N}_0$. Eventually, they dominate the Schawlow–Townes contribution, which decreases with $\bar{N}_0$. This linewidth increase is peculiar to this system, as we remarked before. The relevant interaction energy $\mathcal{W}$ relates both to gain from exciton–exciton scattering and to the quartic term in the interaction Hamiltonian producing the self-phase modulation. In conventional lasers, the physical entities producing gain (atoms) and stimulated emission (photons) are of different nature, so that a large gain exists without any pure photon self-phase modulation. Such a linewidth increase is therefore not observed. Of course, such a self-phase modulation can effectively be produced by inserting a Kerr medium inside the optical cavity of a conventional laser. It has been shown that the phase variance of a coherent state is increased upon passing through a Kerr medium, whereas the number statistics remain Poissonian [212]. The increased phase diffusion is clearly also related to this effect.

In numerical calculations on posts of various sizes, we have shown that a substantial linewidth enhancement of the conventional type can be observed, whereas the effects of the self-modulation term are barely visible. This fact is related to the impossibility of reaching the far-above-threshold condition in the structures investigated. As remarked above, longer lifetimes and/or different materials should allow operation of the laser in this regime, where the new effects should easily become evident.

7.4 Observation of Polariton Oscillation

In Sect. 6.4 we discussed at length the relevance of interface disorder to the dynamics of the microcavity polariton system. We do not have to add any important remarks here for the case of the micropost, as the lateral confinement of the excitons is on a scale much larger than the Bohr radius a_B and the correlation length of the disorder potential. Therefore, the physics does not change in any relevant way for the exciton. The only remark that is in order here is that the etching process will degrade the material quality at the edges of the post, and nonradiative recombination centers and increased disorder will result in this region. If the surface area of the intersection of this region with the quantum well is small, we do not expect a qualitative change of the picture we have presented for the dynamics of microcavity exciton polaritons. This should be the case for relatively large posts with a lateral size of few microns. As for the scattering processes from the confined polaritons to exciton states mediated by disorder, these have been clearly shown to be negligible under appropriate conditions, by four-wave-mixing experiments, in the case of planar cavities [207]. In the micropost case, the lateral confinement raises the energy of the confined lower polariton, and makes this type of scattering more likely. Samples having smaller inhomogeneous broadening than in the planar-cavity case are thus required. We remark that this type of scattering out of the confined polariton states can be effectively taken into account in our model as an additional loss term.

Let us digress to the possibility of observation of micropost polariton lasing in materials other than GaAs. In II–VI material, such as ZnSe- and CdTe-based structures, larger Rabi splittings have been obtained owing to larger exciton oscillator strengths. We also remark that the interface quality of the quantum wells is now comparable to that obtained with GaAs. In particular, similar inhomogeneous linewidths are observed in planar microcavities. A similar ratio l/a_B of the diffusion length to the Bohr radius can thus be expected. What is most interesting is that a larger gain is expected at the critical density at which the Rabi splitting collapses. This maximum-gain density is an increasing function of the value of the Rabi splitting itself, because both the gain and the Rabi-splitting collapse are caused by the same exciton–exciton scattering process. As the gain is proportional to the square of the exciton density, in these materials with large Rabi splittings it can easily be four or more times larger than in GaAs. It is possible that this larger scattering efficiency is sufficient to overcome the shorter radiative lifetimes of the $k = 0$ polaritons of the planar systems which are currently available. The observation of emission nonlinearities with pumping in a II–VI structure reported in [213], interpreted as polariton lasing, could therefore originate from exciton–exciton scattering. The stimulated emission gain up to ∼ 30 and self-oscillation at the bottle-neck polariton state were observed in a CdTe quantum well microcavity [214]. As for the recent observation of lasing in a planar GaAs microcavity reported in [215], this is in contrast to previous experimental results [216], and the parameters of this cavity and experiment need to be clarified. For one thing, the polariton lifetimes in these more recent samples where strong emission nonlinearities were observed were up to five times longer than in the older samples, where the emission nonlinearities followed the collapse of the Rabi splitting. The coherent parametric gain up to ∼ 50 at the bottle-neck polariton state was observed in a GaAs microcavity [217].

It is, finally, instructive to compare the lasing in the micropost system with the stimulated emission of bosonic atoms. The systems show strong analogies. The emission of matter waves, not photons, is stimulated. A cold atomic reservoir provides the atoms, which are scattered into the lowest confined mode of a trap [218]. Many of the proposed gain mechanisms are also based on atomic collisions [218, 219]. However, striking differences are found in the parameters of the two systems. The lifetimes (determined by losses) are in the range of several seconds for trapped atoms, but only tens of picoseconds for polaritons. The trapping energies are tens of nanokelvin for atomic traps, but tens of kelvin in the polariton case. Certainly, the long lifetimes favor the realization of stimulated emission in the atomic system, making even tiny gains sufficient to compensate for losses. Condensates have indeed been obtained, and the problem now centers on the efficient extraction of these atoms from the trap, which corresponds to the introduction of finite mirror reflectivity in a conventional laser cavity. The short lifetimes of the polaritons are appealing for fast technological applications. And although a temperature of few kelvin is technologically unfavorable for applications, it is still nine orders of magniture larger than the small temperatures required in atomic traps [220].

Of different nature is the similarity of the micropost polariton laser to a vertical-cavity surface-emitting laser (VCSEL). This similarity is only structural. VCSELs are small but conventional semiconductor lasers. Population inversion of a fermionic reservoir of electrons and holes drives stimulated emission of the confined *photon* mode of the micropost. However, in the micropost polariton laser the confined polariton which undergoes stimulated emission has an exciton content well over 70%. The excitation densities are also an order of magnitude larger in VCSELs. Although the low operating density and threshold are technologically appealing, a price has to be paid in form of the small operating temperature of the polariton laser, which is severely limited by the material parameters and, in particular, by the Rabi splitting $\hbar\Omega$ and the exciton binding energy E_{B}. A wide bandgap material such as GaN is appealing in this respect: It may open up the possibility of a room-temperature polariton laser.

References

1. S. Haroche and D. Kleppner, Phys. Today, January, 1989, p. 24.
2. E. M. Purcell, Phys. Rev. **69**, 681 (1946).
3. K. H. Drexhage, in *Progress in Optics*, edited by E. Wolf (North-Holland, New York, 1974).
4. P. Goy, J. M. Raimond, M. Gross, and S. Haroche, Phys. Rev. Lett. **50**, 1903 (1983).
5. R. G. Hulet, E. S. Hilfer, and D. Kleppner, Phys. Rev. Lett. **55**, 2137 (1985).
6. W. Jhe, A. Anderson, E. A. Hinds, D. Meschede, L. Moi, and S. Haroche, Phys. Rev. Lett. **58**, 666 (1987).
7. D. J. Heinzen, J. J. Childs, J. F. Thomas, and M. S. Feld, Phys. Rev. Lett. **58**, 1320 (1987).
8. F. De Martini, G. Innocenti, G. R. Jacobovitz, and P. Mataloni, Phys. Rev. Lett. **59**, 2955 (1987).
9. D. J. Heinzen and M. S. Feld, Phys. Rev. Lett. **59**, 2623 (1987).
10. V. Sandoghdar, C. Sukenik, E. Hinds, and S. Haroche, Phys. Rev. Lett. **68**, 3432 (1992).
11. Y. Kaluzny, P. Goy, M. Gross, J. M. Raimond, and S. Haroche, Phys. Rev. Lett. **51**, 1175 (1983).
12. F. Bernardot, P. Nussenveig, M. Brune, J. M. Raimond, and S. Haroche, Europhys. Lett. **17**, 33 (1992).
13. R. J. Thompson, G. Rempe, and H. J. Kimble, Phys. Rev. Lett. **68**, 1132 (1992).
14. G. Rempe, H. Walther, and N. Klein, Phys. Rev. Lett. **58**, 353 (1987).
15. P. Filipowicz, J. Javanainen, and P. Meystre, Phys. Rev. A **34**, 3077 (1986).
16. D. Meschede, H. Walther, and N. Klein, Phys. Rev. Lett. **54**, 551 (1985).
17. M. Brune, J. M, Raimond, P. Goy, L. Davidovich, and S. Haroche, Phys. Rev. Lett. **59**, 1899 (1987).
18. G. Rempe, F. Schmidt-Kaler, and H. Walther, Phys. Rev. Lett. **64**, 2783 (1990).
19. K. An, J. J. Childs, R. R. Dasari, M. S. Feld, and R. George, Phys. Rev. Lett. **73**, 3375 (1994).
20. M. G. Raizen, L. A. Orozco, M. Xiao, T. L. Boyd, and H. J. Kimble, Phys. Rev. Lett. **59**, 198 (1987).
21. E. Hanamura and H. Haug, Phys. Reports **33**, 209 (1977).
22. G. H. Wannier, Phys. Rev. **52**, 191 (1937).
23. L. C. Andreani, in *Confined Electrons and Photons*, edited by E. Burstein, and C. Weisbuch (Plenum, New York, 1995).
24. C. Weisbuch and B. Vinter, *Semiconductor Quantum Structures* (Academic Press, Boston, 1991).
25. J. J. Hopfield, Phys. Rev. **112**, 1555 (1958).

26. F. Bassani and G. Pastori Parravicini, *Electronic States and Optical Transitions in Solids* (Pergamon Press, Oxford, 1975).
27. V. M. Agranovich, Sov. Phys. JETP **37**, 307 (1960).
28. K. Cho and M. Kawata, J. Phys. Soc. Jpn. **54**, 4431 (1985).
29. A. D'Andrea and R. Del Sole, J. Phys. Soc. Jpn. **21**, 1936 (1966).
30. R. C. Miller, D. A. Kleinman, W. T. Tsang, and A. C. Gossard, Phys. Rev. B **24**, 1134 (1981).
31. G. Bastard, E. E. Mendez, L. L. Chang, and E. Esaki, Phys. Rev. B **26**, 1974 (1982).
32. M. Shinada and S. Sugano, Phys. Rev. B **41**, 1413 (1990).
33. R. L. Greene, K. K. Bajaj, and D. E. Phelps Phys. Rev. B **29**, 1807 (1984).
34. L. C. Andreani and A. Pasquarello, Phys. Rev. B **42**, 8928 (1990).
35. B. I. Green, J. Orenstein, and S. Schmitt-Rink, Science **247**, 679 (1989).
36. V. S. Williams, S. Mazumdar, N. R. Armstrong, Z. Z. Ho, and N. Peyghambarian, J. Phys. Chem. **96**, 4500 (1992).
37. L. C. Andreani, F. Tassone, and F. Bassani, Phys. Rev. B **77**, 641 (1991).
38. E. Hanamura, Phys. Rev. B **38**, 1228 (1988)
39. D. S. Citrin, Solid State Commun. **84**, 281 (1992).
40. D. S. Citrin, Phys. Rev. B **47**, 3832 (1993).
41. B. Deveaud, F. Clérot, N. Roy, K. Satzke, B. Sermage, and D. S. Katzer, Phys. Rev. Lett. **67**, 2355 (1991).
42. B. Sermage, B. Deveaud, K. Satzke, F. Clérot, D. Dumas, N. Roy, D. S. Katzer, F. Mollot, R. Planel, M. Berz, and J. L. Oudar, Superl. Microstr. **13**, 271 (1993).
43. D. S. Citrin, Comments Condens. Matt. Phys. **16**, 263 (1993).
44. R. E. Slusher and C. Weisbuch, Solid State Commun. **92**, 149 (1994).
45. E. Yablonovitch, Phys. Rev. Lett. **58**, 2059 (1987).
46. G. Björk, Y. Yamamoto, and H. Heitmann, in *Confined Electrons and Photons*, edited by E. Burstein and C. Weisbuch (Plenum, New York, 1995).
47. A. Yariv, *Quantum Electronics* (Wiley, New York, 1989).
48. T. Baba, T. Hamano, F. Koyama, and K. Iga, IEEE J. Quant. Electr. **27**, 1347 (1991).
49. Y. Yamamoto, S. Machida, K. Igata, and Y. Horikoshi, in *Coherence and Quantum Optics VI*, edited by J. H. Eberly, L. Mandel, and E. Wolf (Plenum, New York, 1989), p. 1243.
50. C. Weisbuch, M. Nishioka, A. Ishikawa, and Y. Arakawa, Phys. Rev. Lett. **69**, 3314 (1992).
51. Y. Zhu, D. J. Gauthier, S. E. Morin, Q. Wu, H. J. Carmichael, and T. W. Mossberg, Phys. Rev. Lett. **64**, 2499 (1990).
52. D. S. Citrin, IEEE J. Quant. Electr. **30**, 997 (1994).
53. F. Tassone, F. Bassani, and L. C. Andreani, Phys. Rev. B **45**, 6023 (1992).
54. L. C. Andreani, Phys. Lett. A **192**, 99 (1994).
55. R. Houdrè, R. P. Stanley, U. Oesterle, M. Ilegems, C. Weisbuch, and Y. Arakawa, J. Phys. IV **3**, 51 (1993).
56. V. Savona, L. C. Andreani, P. Schwendimann, and A. Quattropani, Solid State Commun. **93**, 733 (1995).
57. F. Bassani, F. Ruggiero, and A. Quattropani, Nuovo Cimento D **7**, 700 (1986).
58. S. Pau, G. Björk, J. Jacobson, H. Cao, and Y. Yamamoto, Phys. Rev. B **51**, 14437 (1995).
59. V. Savona, Z. Hradil, A. Quattropani, and P. Schwendimann, Phys. Rev. B **49**, 8774 (1994).
60. P. Meystre and M. Sargent III, *Quantum Statistical Properties of Radiation* (Wiley, New York, 1973).

61. R. Houdré, C. Weisbuch, R. P. Stanley, U. Oesterle, P. Pellandini, and M. Ilegems, Phys. Rev. Lett. **73**, 2043 (1994).
62. T. B. Norris, J.-K. Rhee, C.-Y. Sung, Y. Arakawa, M. Nishioka, and C. Weisbuch, Phys. Rev. B **50**, 14663 (1994).
63. J. Jacobson, S. Pau, H. Cao, G. Björk, and Y. Yamamoto, Phys. Rev. A **51**, 2542 (1995).
64. H. Wang, J. Shah, T. C. Damen, W. Y. Jan, J. E. Cunningham, M. Hong, and J. P. Mannaerts, Phys. Rev. B **51**, 14713 (1995).
65. R. Houdré, R. P. Stanley, U. Oesterle, M. Ilegems, and C. Weisbuch, Phys. Rev. B **49**, 16761 (1994).
66. T. R. Nelson, Jr., J. P. Prineas, G. Khitrova, H. M. Gibbs, J. D. Berger, and E. K. Lindmark, Appl. Phys. Lett. **69**, 3031 (1996).
67. P. Kelkar, V. Kozlov, H. Jeon, A. V Nurmikko, C.-C. Chu, D. C. Brillo, J. Han, C. G. Hua, and R. L. Gunshor, Phys. Rev. B **52**, R5491 (1995).
68. S. Jiang, S. Machida, Y. Takiguchi, H. Cao, and Y. Yamamoto, Appl. Phys. Lett. **73**, 3031 (1998).
69. H. Cao, S. Jiang, S. Machida, Y. Takiguchi, and Y. Yamamoto, Appl. Phys. Lett. **71**, 1461 (1997).
70. J. H. Eberly, N. B. Narozhny, and J. J. Sanchez-Mondragon, Phys. Rev. Lett. **44**, 1323 (1980).
71. H. Cao, J. Jacobson, G. Björk, S. Pau, and Y. Yamamoto, Appl. Phys. Lett. **66**, 1107 (1995).
72. C. Weisbuch, R. Dingle, A. C. Gossard, and W. Wiegmann, Solid State Commun. **38**, 709 (1981).
73. S. Pau, G. Björk, H. Cao, E. Hanamura, and Y.Yamamoto, Solid State Commun. **98**, 781 (1996).
74. H. Cao, S. Pau, Y.Yamamoto, and G. Björk, Phys. Rev. B **54**, 8083 (1996).
75. B. I. Halperin, Phys. Rev. **139**, A104 (1965).
76. D. M. Whittaker, P. Kinsler, T. A. Fisher, M. S. Skolnick, A. Armitage, A. M. Afshar, M. D. Sturge, and J. S. Roberts, Phys. Rev. Lett. **77**, 4792 (1996).
77. V. Savona, C. Piermarocchi, A. Quattropani, F. Tassone, and P. Schwendimann, Phys. Rev. Lett. **78**, 4470 (1997).
78. S. Pau, G. Björk, J. M. Jacobson, H. Cao, and Y. Yamamoto, Phys. Rev. B **51**, 7090 (1995).
79. M. V. Klein, IEEE J. Quant. Electr. **QE-22**, 1760 (1986).
80. W. C. Tait and R. L. Weiher, Phys. Rev. **178**, 1404 (1969).
81. A. Quattropani, L. C. Andreani, and F. Bassani, Nuovo Cimento D **7**, 55 (1986).
82. Y. Shinozuka and M. Matsuura, Phys. Rev. B **29**, 3717 (1984).
83. T. Takagahara, Phys. Rev. B **31**, 6552 (1985).
84. B. K. Ridley, *Quantum Processes in Semiconductors*, (Oxford University Press, Oxford, 1993).
85. B. R. Nag, *Theory of Electrical Transport in Semiconductors* (Pergamon, Oxford, 1972).
86. J. Lee, E. S. Koteles, and M. O. Vassell, Phys. Rev. B **33**, 5512 (1986).
87. R. P. Stanley, R. Houdré, C. Weisbuch, U. Oesterle, and M. Ilegems, Phys. Rev. B **53**, 10995 (1996).
88. F. Tassone, C. Piermarocchi, V. Savona, and A. Quattropani, Phys. Rev. B **53**, R7642 (1996).
89. S. Pau, G. Björk, H. Cao, R. Huang, Y.Yamamoto, and R. P. Stanley, Phys. Rev. B **55**, R1942 (1997).
90. A. Fainstein, B. Jusserand, and V. Thierry-Mieg, Phys. Rev. Lett. **75**, 3764 (1995).

91. A. J. Shields, M. P. Chamberlain, M. Cardona, and K. Eberl, Phys. Rev. B 51, 17728 (1995); J. E. Zucker, A. Pinczuk, D. S. Chemla, and A. C. Gossard, Phys. Rev. B 35, 2892 (1987).

92. B. Jusserand and M. Cardona, in *Light Scattering in Solids*, edited by M. Cardona and G. Güntherodt (Springer, Berlin, Heidelberg, 1989), Vol. 5.

93. J. A. Kash, S. S. Jha, and J. C. Tsang, Phys. Rev. Lett. 58, 1869 (1987).

94. D. A. B. Miller, D. S. Chemla, T. C. Damen, A. C. Gossard, W. Wiegmann, T. H. Wood, and C. Burrus, Phys. Rev. B 32, 1043 (1985).

95. T. A. Fisher, A. M. Afshar, D. M. Whittaker, and M. S. Skolnick, Phys. Rev. B 51, 2600 (1995).

96. T. A. Fisher, A. M. Afshar, M. S. Skolnick, and D. M. Whittaker, Phys. Rev. B 53, R10469 (1996).

97. J. D. Berger, O. Lyngnes, H. M. Gibbs, G. Khitrova, T. R. Nelson, E. K. Lindmark, A. V. Kanokin, M. A. Kaliteevski, and V. V. Zapasskii, Phys. Rev. B 54, 1975 (1996).

98. J. Tignon, P. Voisin, D. Delalande, M. Voos, R. Houdrè, U. Oesterle, and R. P. Stanley, Phys. Rev. Lett. 74, 3967 (1995).

99. S. Jiang, S. Machida, Y. Takiguchi, H. Cao, and Y. Yamamoto, Opt. Commun. 145, 91 (1998).

100. S. Schmitt-Rink, D. S. Chemla, and D. A. B. Miller, Phys. Rev. B 32, 6601 (1985).

101. C. Cohen-Tannoudji, J. Dupont-Roc, and G. Grynberg, *Atom-Photon Interactions* (Wiley, New York, 1993).

102. D. C. Reynolds, K. K. Bajaj, C. E. Stutz, and R. L. Jones, Phys, Rev. B 40, 3340 (1989).

103. R. Houdré, J. L. Gibernon, P. Pellandini, R. P. Stanley, U. Oesterle, C. Weisbuch, J. O'Gorman, B. Roycroft, and M. Ilegems, Phys. Rev. B 52, 7810 (1995).

104. F. Jahnke, M. Kira, S. W. Koch, G. Khitrova, E. K. Lindmark, T. R. Nelson, D. V. Wick, J. D. Berger, O. Lyngnes, H. M. Gibbs, and K. Tai, Phys. Rev. Lett. 77, 5257 (1996).

105. J.-K. Rhee, D. S. Citrin, T. B. Norris, Y. Arakawa, and M. Nishioka, Solid State Commun. 97, 941 (1996).

106. H. Wang, J. Shah, T. C. Damon, and L. N. Pfeiffer, Solid State Commun. 91, 869 (1994).

107. J.-Y. Bigot, A. Daunois, J. Oberle, and J. -C. Merle, Phys. Rev. Lett. 71, 1820 (1993).

108. M. K-Gonokami, S. Inouye, H. Suzuura, M. Shirane, and R. Shimano, Phys. Rev. Lett. 79, 1341 (1997).

109. A. M. Fox, D. A. B. Miller, G. Livescu, J. E. Cunningham, J. E. Henry, and W. Y. Jan, Phys. Rev. B 42, 1841 (1990).

110. H. Schneider, J. Wagner, and K. Ploog, Phys. Rev. B 48, 11051 (1993).

111. R. Ferreira, P. Rolland, P. Roussignol, C. Delalande, and A. Vinattieri, Phys. Rev. B 45, 11782 (1992).

112. C. Delalande, in *Resonant Tunneling in Semiconductors*, edited by L. L. Chang, E. E. Mendez, and C. Tejedor (Plenum, New York, 1991).

113. I. Lawrence, S. Haacke, H. Mariette, W. W. Rühle, H. Ulmer-Tuffigo, J. Cibert, and G. Feuillet, Phys. Rev. Lett. 73, 2131 (1994).

114. H. Cao, G. Klimovitch, G. Björk, and Y. Yamamoto, Phys. Rev. Lett. 75, 1146 (1995).

115. J. Bardeen, Phys. Rev. Lett. 6, 57 (1961).

116. G. Garcia Calderón, in *Resonant Tunneling in Semiconductors*, edited by L. L. Chang, E. E. Mendez, and C. Tejedor (Plenum, New York, 1991).

117. P. Guéret and C. Rossel, in *Resonant Tunneling in Semiconductors*, edited by L. L. Chang, E. E. Mendez, and C. Tejedor (Plenum, New York, 1991).
118. R. Wessel, in *Resonant Tunneling in Semiconductors*, edited by L. L. Chang, E. E. Mendez, and C. Tejedor (Plenum, New York, 1991).
119. B. Vinter and F. Chevoir, in *Resonant Tunneling in Semiconductors*, edited by L. L. Chang, E. E. Mendez, and C. Tejedor (Plenum, New York, 1991).
120. K. Leo, in *Optics of Semiconductor Nanostructures*, edited by F. Henneberger, S. Schmitt-Rink, and E. O. Göbel (Akademia, Berlin, 1993).
121. G. Fishman, Solid State Commun. **27**, 1097 (1978).
122. R. J. Elliot, Phys. Rev. **124**, 340 (1961).
123. Y. Yamamoto and R. E. Slusher, Phys. Today **46**, 66 (June 1993).
124. G. Björk, J. Jacobson, S. Pau, and Y. Yamamoto, Phys. Rev. B **50**, 17336 (1994).
125. A. Celeste, L. A. Cury, J. C. Portal, M. Allovon, K. D. Maude, L. Eaves, M. Davies, M. Heath, and M. Maldonado, in *Resonant Tunneling in Semiconductors*, edited by L. L. Chang, E. E. Mendez, and C. Tejedor (Plenum, New York, 1991).
126. E. V. Anda and F. Flores, in *Resonant Tunneling in Semiconductors*, edited by L. L. Chang, E. E. Mendez, and C. Tejedor (Plenum, New York, 1991).
127. D. C. Rogers, J. Singleton, R. J. Nicholas, C. T. Foxon, and K. Woodbridge, Phys. Rev. B **34**, 4002 (1986).
128. P. Dawson, K. J. Moore, G. Duggan, H. I. Ralph, and C. T. Foxon, Phys. Rev. B **34**, 6007 (1987).
129. E. S. Koteles and J. Y. Chi, Phys. Rev. B **37**, 6332 (1988).
130. D. A. B. Miller, in *Confined Electrons and Photons*, edited by E. Burstein and C. Weisbuch (Plenum, New York, 1995).
131. J.-P. Reithmaier, R. Hoger, and H. Riechert, Phys. Rev. B **43**, 4933 (1991).
132. C. Comte and P. Nozieres, J. Physique **43**, 1069 (1982)
133. V. N. Popov and V. S. Yarunin, *Collective Effects in Quantum Statistics of Radiation and Matter*, (Kluwer Academic, Dordrecht, 1988), Chaps. 2 and 3.
134. L. V. Keldysh and Yu. V. Kopaev, Sov. Phys. Solid State **6**, 2219 (1965).
135. L. Belkhir and M. Randeria, Phys. Rev. B **49**, 6829 (1994).
136. A. Bardasis and J. R. Schrieffer Phys. Rev. **121**, 1050 (1961).
137. R. Cote and A. Griffin, Phys. Rev. B **37**, 4539 (1988).
138. G. Klimovitch, *PhD thesis*, Stanford University, California (1999).
139. G. Klimovitch, G. Björk, H. Cao, and Y. Yamamoto, Phys. Rev. B **55**, 7078 (1997).
140. J. M. Gerard, D. Barrier, J. Y. Marzin, R. Kuszelewicz, L. Manin, E. Costard, V. Thierry-Mieg, and T. Rivera, Appl. Phys. Lett. **69**, 449 (1996).
141. J. P. Reithmaier, M. Rohner, H. Zull, F. Schafer, and A. Forchel, Phys. Rev. Lett. **78**, 378 (1997).
142. M. H. Anderson, J. R. Ensher, M. R. Matthews, C. E. Wieman, and E. A. Cornell, Science **269**, 198 (1995).
143. C. C. Bradley, C. A. Sackett, J. J. Tollett, and R. G. Hulet, Phys. Rev. Lett. **75**, 1687 (1995).
144. K. B. Davis, M.-O. Mewes, M. R. Andrews, N. J. van Druten, D. S. Durfee, D. M. Kurn, and W. Keterle, Phys. Rev. Lett. **75**, 3969 (1995).
145. P. Nozieres, Physica **117 & 118B**, 16 (1983).
146. V. B. Timofeev, V. D. Kulakovskii, and I. V. Kukushkin, Physica **117 & 118B**, 327 (1983).
147. N. Peyghambarian, L. L. Chase, and A. Mysyrowicz, Phys. Rev. B **27**, 2325 (1983).
148. D. Snoke, J. P. Wolfe, and A. Mysyrowicz, Phys. Rev. Lett. **59**, 827 (1987).

149. V. T. Agekyan, Phys. Stat. Sol. A **43**, 11 (1977).
150. J. L. Lin and J. P. Wolfe, Phys. Rev. Lett. **71**, 1222 (1993).
151. Y. E. Lozovik and V. I. Yudson, JETP Lett. **22**, 274 (1975).
152. S. I. Shevchenko, Sov. J. Low Temp. Phys. **2**, 251 (1976).
153. T. Fukuzawa, S. S. Kano, T. K. Gustafson, and T. Ogawa, Surf. Sci. **228**, 482 (1990).
154. L. V. Butov, A. Zrenner, G. Abstreiter, G. Böhm, and G. Weimann, Phys. Rev. Lett. **73**, 304 (1994).
155. H. M. Wiseman and M. J. Collett, Phys. Lett. A **202**, 246 (1995).
156. M. Holland, K. Burnett, C. Gardiner, J. I. Cirac, and P. Zoller, Phys. Rev. A **54**, R1757 (1996).
157. A. Imamoğlu and R. J. Ram, Phys. Lett. A **214**, 193 (1996).
158. A. Griffiu, D. W. Snoke, and S. Stringari, eds., *Bose-Einstein Condensation* (Cambridge University Press, New York, 1995).
159. L. V. Keldysh and A. N. Kozlov, Sov. Phys. JETP **27**, 521 (1968).
160. E. Hanamura and H. Haug, Phys. Rep. **33**, 209 (1977).
161. C. Comte and P. Nozieres, J. Physique **43**, 1069 (1982).
162. A. Imamoğlu, R. J. Ram, S. Pau, and Y. Yamamoto, Phys. Rev. A **53**, 4250 (1996).
163. W. W. Chow, S. W. Koch, and M. Sargant III, *Semiconductor Laser Physics* (Springer, Berlin, Heidelberg, 1994).
164. M. E. Flatte, E. Runge, and H. Ehrenreich, Appl. Phys. Lett. **66**, 1313 (1995).
165. S. Pau, J. Jacobson, G. Björk, and Y. Yamamoto, J. Opt. Soc. Am. B **13**, 1078 (1996).
166. O. Kocharovskaya and Ya I. Khanin, JETP Lett. **48**, 630 (1988).
167. S. E. Harris, Phys. Rev. Lett. **62**, 1033 (1989).
168. M. O. Scully, S.-Y. Zhu, and A. Gavrielides, Phys. Rev. Lett. **62**, 2813 (1989).
169. A. Imamoğlu, Phys. Rev. A **40**, 2835 (1989).
170. S. Pau, H. Cao, J. Jacobson, G. Björk, Y. Yamamoto, and A. Imamoğlu, Phys. Rev. A **54**, R1789 (1996).
171. Y. Yamamoto, S. Machida, K. Igeta, and G. Björk, in *Coherence, Amplification, and Quantum Effects in Semiconductor Lasers*, ed. Y. Yamamoto (Wiley, New York, 1991).
172. J. P. Zhang, D. Y. Chu, S. L. Wu, W. G. Bi, R. C. Tiberio, C. W. Tu, and S. T. Ho, IEEE Photon. Tech. Lett. **8**, 968 (1996).
173. P. Kelkar, V. Kozlov, H. Jeon, A. V. Nurmiko, C. C. Chu, D. C. Grillo, J. Han, C. G. Hua, and R. L. Gunshor, Phys. Rev. B **52**, R5491 (1996).
174. M. Kira, F. Jahnke, and S.W. Koch, Phys. Rev. Lett. **81**, 3263 (1998).
175. C. Piermarocchi, V. Savona, A. Quattropani, P. Schwendimann, and F. Tassone, Phys. Stat. Sol. A **164**, 221 (1997).
176. P. Danielewicz, Ann. Phys. **152**, 239 (1984).
177. N. N. Bogolyubov, V. V. Tolmechev, and D. V. Shirkov, in *A New Method in the Theory of Superconductivity* (New York, Consultants Bureau, 1959).
178. V. Savona, L. C. Andreani, A. Quattropani, and P. Schwendimann, Solid State Commun. **93**, 733 (1995).
179. V. Savona, F. Tassone, C. Piermarocchi, A. Quattropani and P. Schwendimann, Phys. Rev. B **53**, 13051 (1996).
180. C. Piermarocchi, F. Tassone, V. Savona, A. Quattropani and P. Schwendimann, Phys. Rev. B **53**, 15834 (1996).
181. G. Wentzel, Phys. Rev. **108**, 1593 (1957).
182. T. Usui, Progr. Theor. Phys. **23**, 787 (1960).
183. E. Hanamura and H. Hang, Phys. Reports **33**, 203 (1977).

184. S. K. Ma, *Modern Theory of Critical Phenomena* (W. A. Benjamin, Reading, MA, 1976).
185. F. Čulik, Czech. J. Phys. **16**, 194 (1966).
186. A. I. Bobryesheva, M. F. Miglei, and M. I. Shmiglyuk, Phys. Stat. Sol. B **53**, 71 (1972); A. I. Bobryesheva, V. T. Zyukov, and S. I. Beryl, Phys. Stat. Sol. B **101**, 69 (1980).
187. C. Ciuti, V. Savona, C. Piermarochhi, A. Quattropani, and P. Schwendimann, Phys. Rev. B **58**, 7926 (1998).
188. F. Tassone and Y. Yamamoto, Phys. Rev. B **59**, 10830 (1999).
189. D.W. Snoke and J.P. Wolfe, Phys. Rev. B **39**, 4030 (1989).
190. L. C. Andreani, F. Tassone, and F. Bassani, Solid State Commun. **77**, 641 (1991).
191. F. Tassone, C. Piermarocchi, V. Savona, A. Quattropani, and P. Schwendimann, Phys. Rev. B **56**, 7554 (1997).
192. D. M. Whittaker, P. Kinsler, T. A. Fisher, M. S. Skolnick, A. Armitage, A. M. Afshar, M. D. Sturge, and J. S. Roberts, Phys. Rev. Lett. **77**, 4792 (1996).
193. V. Savona, C. Piermarocchi, A. Quattropani, F. Tassone, and P. Schwendimann, Phys. Rev. Lett. **78**, 447 (1997).
194. D. M. Whittaker, Phys. Rev. Lett. **80**, 4791 (1998).
195. C. Ell, J. Prineas, T. R. Nelson, Jr., S. Park, H. M. Gibbs, G. Khitrova, S. W. Koch, and R. Houdré, Phys. Rev. Lett. **80**, 4795 (1998).
196. R. P. Stanley, S. Pau, U. Oesterle, R. Houdré, and M. Ilegems, Phys. Rev. B **55**, R4867 (1997).
197. H. Cao, S. Pan, J. M. Jacobson, G. Björk, Y. Yamamoto, and A. Imamoglu, Phys. Rev. A, **55**, 4632 (1997).
198. M. Kira, F. Jahnke, S. W. Koch, J. D. Berger, D. V. Wick, T. R. Nelson, Jr., G. Khitrova, and H. M. Gibbs, Phys. Rev. Lett. **79**, 5170 (1997).
199. F. Tassone, C. Piermarocchi, V. Savona, A. Quattropani, and P. Schwendimann, Phys. Rev. B **53**, R 7642 (1996).
200. J. Wainstain, G. Cassabois, Ph. Roussignol, C. Delalande, M. Voos, F. Tassone, R. Houdré, R. P. Stanley, and U. Oesterle, Superlattices and Microstructures **22**, 389 (1997).
201. M. Kuwata-Gonokami, S. Inouye, H. Suzuura, M. Shirane, and R. Shimano, Phys. Rev. Lett. **79**, 1341 (1997).
202. R. Houdré, J. L. Gibernon, P. Pellandini, R. P. Stanley, U. Oesterle, C. Weisbuch, J. O'Gorman, B. Roycroft, and M. Ilegems, Phys. Rev. B **52**, 7810 (1995).
203. F. Jahnke, M. Kira, S. W. Koch, G. Khitrova, E. K. Lindmark, T. R. Nelson, Jr., D. V. Wick, J. D. Berger, O. Lyngnes, H. M. Gibbs, and K. Tai, Phys. Rev. Lett. **77**, 5257 (1996).
204. F. Tassone, R. Huang, and Y. Yamamoto, e-print cond-mat/9808157.
205. R. Huang, F. Tassone, and Y. Yamamoto, Phys. Rev. B **61**, R7854 (2000).
206. R. Huang, F. Tassone, and Y. Yamamoto, *Microelectronic Engineering* **47**, 325 (1999).
207. J. J. Baumberg, A. Armitage, M. S. Skolnick, and J. S. Roberts, Phys. Rev. Lett. **81**, 661 (1998).
208. W. H. Louisell, *Quantum Statistical Properties of Radiation*, (Wiley, New York, 1973).
209. N. N. Bogolyubov, V. V. Tolmachev, and D. V. Shirkov, *A New Method in the Theory of Superconductivity*, (Consultants Bureau, New York, 1959).
210. F. Tassone and Y. Yamamoto, submitted to Phys. Rev. A.
211. Y. Yamamoto, S. Machida, and O. Nilsson, Phys. Rev. A **34**, 4025 (1986).
212. M. Kitagawa and Y. Yamamoto, Phys. Rev. A **34**, 3974 (1986).

213. Le Si Dang, D. Heger, R. André, F. Bœuf, and R. Romenstein, Phys. Rev. Lett. **81**, 3920 (1998).
214. R. Huang, Ph.D. dissertation thesis (Stanford University, 2000).
215. P. Sennellart and J. Bloch, Phys. Rev. Lett. **82**, 1233 (1999).
216. H. Cao, S. Pau, J. M. Jacobson, G. Björk, Y. Yamamoto, and A. Imamoğlu, Phys. Rev. A **55**, 4632 (1997); X. Fan, H. Wang, H. Q. Hou, and B. E. Hammons, Phys. Rev. A **56**, 3236 (1997); Xudong Fan, H. Wang, H. Q. Hou, and B. E. Hammons, Phys. Rev. B **56**, 15256 (1997).
217. P. Savridis, J. J. Baumberg, R. M. Stevenson, M. S. Skolnick, D. M. Whittaker, and J. S. Roberts, Phys. Rev. Lett. **84**, 1547 (2000); Y. Yamamoto, Nature **405**, 630 (2000).
218. H. M. Wiseman and M. J. Collet, Phys. Lett. A **201**, 246 (1995); R. J. C. Spreeuw, T. Pfau, U. Janicke, and M. Wilkens, Europhys. Lett. **32**, 469 (1995).
219. A. M. Guzman, M. Moore, and P. Meystre, Phys. Rev. A **53**, 977 (1996).
220. M. H. Anderson, R. J. Ensher, M. R. Matthews, C. E. Wiemann, and E. A. Cornell, Science **269**, 198 (1995).

Index

Printing: Mercedes-Druck, Berlin
Binding: Stürtz AG, Würzburg

Springer Tracts in Modern Physics